Springer Finance

Springer
Berlin
Heidelberg
New York
Hong Kong
London
Milan
Paris
Tokyo

Springer Finance

Springer Finance is a new programme of books aimed at students, academics and practitioners working on increasingly technical approaches to the analysis of financial markets. It aims to cover a variety of topics, not only mathematical finance but foreign exchanges, term structure, risk management, portfolio theory, equity derivatives, and financial economics.

Credit Risk Valuation:
Risk-Neutral Valuation:
Pricing and Hedging of Finance Derivatives
Bingham, N. H. and Kiesel, R.
ISBN 1-85233-001-5 (1998)

Visual Explorations in Finance
with Self-Organizing Maps
Deboeck, G. and Kohonen, T. (Editors)
ISBN 3-540-76266-3 (1998)

Mathematical Models of Financial Derivatives
Kwok, Y.-K.
ISBN 3-981-3083-25-5 (1998)

Mathematics of Financial Markets
Elliott, R. J. and Kopp, P. E.
ISBN 0-387-98533-0 (1999)

Efficient Methods
for Valuing Interest Rate Derivatives
Pelsser, A.
ISBN 1-85233-304-9 (2000)

Methods, Models and Applications
Ammann, M.
ISBN 3-540-67805-0 (2001)

Credit Risk: Modelling, Valuation and Hedging
Bielecki, T. R. and Rutkowski, M.
ISBN 3-540-67593-0 (2001)

Mathematical Finance – Bachelier Congress 2000
– Selected Papers from the First World Congress
of the Bachelier Finance Society,
held in Paris, June 29–July 1, 2000
Geman, H., Madan, D. S., Pliska R. and Vorst, T. (Editors)
ISBN 3-540-67781-X (2001)

Exponential Functionals of Brownian Motion
and Related Processes
Yor, M.
ISBN 3-540-65943-9 (2001)

Financial Markets Theory:
Equilibrium, Efficiency and Information
Barucci, E.
ISBN 3-85233-469-X (2003)

Financial Markets in Continuous Time
Dana, R.-A. and Jeanblanc, M.
ISBN 3-540-41722-9 (2003)

Weak, Convergence of Financial Markets
Prigent, J.-L.
ISBN 3-540-4233-8 (2003)

Incomplete Information and Heterogenous
Beliefs in Continuous-time Finance
Ziegler, A.
ISBN 3-540-00344-4 (2003)

Stochastic Calculus Models for Finance:
Volume 1: The Binominal Assett Pricing Model
Shreve, S. E.
ISBN 3-540-40101-6 (2004)

Irrational Exuberance Reconsidered:
The Cross Section of Stock Returns
Külpmann, M.
ISBN 3-540-14007-7 (2004)

Credit Risk Pricing Models: Theory and Practice
Schmid, B.
ISBN 3-54040466-X (2004)

A Game Theory Analysis of Options
Ziegler, A.
ISBN 3-540-20668-X (2004)

Asset Pricing: Modeling and Estimation
Kellerhals, B. P.
ISBN 3-540-20853-4 (2004)

B. Philipp Kellerhals

Asset Pricing

Modeling and Estimation

Second Edition
with 10 Figures
and 30 Tables

Springer

Dr. B. Philipp Kellerhals
dit | Allianz Dresdner Asset Management
Mainzer Landstrasse 11–13
60329 Frankfurt am Main
Germany

Mathematics Subject Classification (2000):
91B28

Originally published as volume 506 in the series "Lecture Notes in Economics and Mathematical Systems"

ISBN 3-540-20853-4 Springer-Verlag Berlin Heidelberg New York

Bibliographic information published by Die Deutsche Bibliothek
Die Deutsche Bibliothek lists this publication in the Deutsche Nationalbibliografie; detailed bibliographic data available in the internet at *http.//dnb.ddb.de*

Springer-Verlag is a part of Springer Science+Business Media
springeronline.com

Cover design: design & production, Heidelberg

SPIN 10982570 42/3130 – 5 4 3 2 1 0 – Printed on acid-free paper

To Gerhard and Susanne.

Preface

The modern field of asset pricing asks for sound pricing models grounded on the theory of financial economics à la Ingersoll (1987) as well as for accurate estimation techniques à la Hamilton (1994b) when it comes to empirical inferences of the specified model. The idea behind this book on hand is to provide the reader with a canonical framework that shows how to bridge the gap between the continuous-time pricing practice in financial engineering and the capital market data inevitably only available at discrete time intervals.

Three major financial markets are to be examined for which we select the equity market, the bond market, and the electricity market. In each market we derive new valuation models to price selected financial instruments in continuous-time. The decision criterium for choosing a *continuous-time modeling framework* is the richness of the stochastic theory available for continuous-time processes with Merton's pioneering contributions to financial economics, collected in Merton (1992). The continuous-time framework, reviewed and assessed by Sundaresan (2000), allows us to obtain analytical pricing formulae that would be unavailable in a discrete time setting. However, at the time of implementing the derived theoretical pricing models on market data, that is necessarily sampled at discrete time intervals, we work with so-called exact discrete time equivalents à la Bergstrom (1984). We show how to conveniently work within a state space framework which we derive in a general setting as well as explicitly for each of the three applications. Thereupon we perform empirical maximum likelihood inferences implementing linear and extended specifications of Kalman filter algorithms. The *Kalman filter framework* was originally developed by Kalman (1960) and Kalman and Bucy (1961) for applications in aerospace engineering (part of the astronautical guidance system of the Apollo project) and has successfully been used in electrical engineering. Only in the last decade, marked by Harvey (1989), Kalman filters have found ground as econometric tool for economic and financial estimation problems.

The technique of Kalman filtering generally leads to more efficient parameter estimates since it imposes greater theoretical modeling restrictions on the data than other commonly used methods. Given a hypothesized financial pricing model, the estimation procedure of Kalman filtering imposes cross-sectional as well as time-series restrictions, making it a true maximum likelihood method. We use different types of Kalman filter algorithms, which we introduce in chapter 2 of this book, to empirically estimate current pricing problems in the field of financial economics.

The *selected pricing issues* within the chosen financial markets are presented in the three distinct parts II, III, and IV of this book:

In part II we investigate the *pricing of equities.* Within this asset class we specialize in the valuation of closed-end funds which share distinct financial characteristics of common joint-stock companies and mutual fund investment companies. In building our stochastic pricing model we especially draw our attention on the market anomaly that the market value of closed-end funds determined on organized exchanges differs dynamically over time from the reported value on their underlying investment portfolios by a discount or a premium. In light of this anomaly we develop a two-factor valuation model of closed-end funds that takes into account both the price risk of the funds as well as the risk associated with altering discounts. Based on a state space formulation of the pricing model we estimate the relevant model parameters using a Kalman filter framework. Having validated our pricing model on a sample of emerging market closed-end funds traded on the New York Stock Exchange, we further assess the quality of the developed model for investor applications. There, we test the forecasting power of the valuation model to predict closed-end fund market prices and implement portfolio strategies using trading rules. The results of these applications indicate that the pricing model generates valuable investment information.

For *pricing fixed-income securities* we present a new dynamic term structure model in part III which belongs to the affine term structure models. Motivated by the incomplete market situation of fixed-income instruments we propose a pricing model that shifts from modeling static tastes to beliefs of the dynamic behavior of tastes. Besides a stochastic short interest rate the proposed two-factor term structure model allows for a stochastic behavior of the market price of interest rate risk. Given the theoretical term structure model we derive a closed-form expression for valuing discount bonds and examine the implied term structures of interest rates and volatilities as initial characteristic results. We further show how to apply the term structure model in a consistent framework to manage interest rate risk and price interest rate derivative securities. There we start with the concept of immunization using factor duration and price bond options, swap contracts, and interest rate caps and floors as the most relevant derivative contracts in risk management. Using US market data on Treasury Securities, LIBOR rates, and swap rates we implement the theoretical pricing model based on a corresponding state space form. Linear and extended Kalman filter routines give us the possibility of estimating the model as well as extracting the time-series of the underlying state variables.

In part IV we present a continuous-time pricing model for the valuation of *electricity derivatives.* From investigating more mature commodity markets than the electricity market, we gather information on the price behavior of commodities and pricing models for commodity futures. For electricity prices we especially examine the most relevant idiosyncratic price patterns which are considered to be their seasonality, mean-reversion, and extreme volatility. From properly chosen model assumptions, especially capturing the unique characteristic of non-storability of electricity and the marked volatility in electricity prices being both high and variable over time, we build an appropriate valuation model to price electricity forwards. The proposed stochastic volatility pricing model yields a closed-form solution to electricity forwards using risk neutral pricing techniques. Next, we empirically adapt the theoretical pricing model in state space form dealing with state variables that follow a non-Gaussian distribution; thereupon we implement an extended Kalman filter algorithm to estimate the model by means of maximum likelihood. Finally, we report empirical results of the pricing model based on electricity data from the largely deregulated Californian power market.

Each of these three application parts can be read independently and contains an introductory survey chapter which motivates the presented research and clears its context in the respective literature. Furthermore, each part concludes with a summary chapter where we finally present our findings. The asset pricing framework common to all three modeling and estimation applications is comprehensively described in part I of this book.

Although this preface is modified somewhat from that of the first edition, some additional remarks will clarify how the second edition differs from the first, Kellerhals (2001). Chapters 3, 4, and 5 from the first edition are merged into one chapter, Estimation Principles. The references have been thoroughly updated, and nontrivial changes and improvements have been made throughout the chapters. Recent research contributions in the three covered fields of equities with closed-end funds, fixed-income products, and electricity derivatives are integrated in the text. The newly incorporated chapter Financial Modeling especially provides a comprehensive treatment of the arbitrage pricing theory in continuous-time as formidable dynamic asset pricing background for the applications presented in parts II, III, and IV of this book. Finally, I am grateful to have had the chance to extensively work in the field of dynamic asset pricing theory as well as on aspects of econometric modeling, empirical estimation and calibration of applied financial models as a researcher and lecturer in the Department of Finance at the Eberhard-Karls-University Tübingen, Germany, in particular with my academic supervisor Rainer Schöbel. The undergone research presented in the second edition in continuation of the first edition extensively profited from seminars, lectures, suggestions, and discussions from and with Yacine Ait-Sahaliah, Günter Bamberg, Michael Brennan, Wolfgang Bühler, Tarun Chordia, Roland Demmel, Gordon Gemmill, Stephan Heilig, Herbert Heyer, Olaf Korn, Roman Liesenfeld, Hartmut Nagel, Werner Neus, Ariane Reiss, Gerd Ronning, Christian Schlag, Manfred Stadler, Marliese Uhrig-Homburg, Nick Webber, and Jianwei Zhu. I rededicate this book to my parents.

Frankfurt am Main,
January 2004

B. Philipp Kellerhals

Contents

Part I Asset Pricing Framework

Part II Pricing Equities

Part I

Asset Pricing Framework

1

Financial Modeling

Typical financial pricing models in continuous time are built upon hypothesized *stochastic processes* to assess the models' dynamics over time and the underlying economy. A common and convenient method to describe the evolution of the modeled state variables is to specify their probabilistic behavior via Itô processes. Here, the basic building block is the Brownian motion which we examine in this chapter accompanied by the necessary stochastic calculus to set up appropriate stochastic differential equations. The final objective is to come close to finding the real data generating process behind the modeled capital market. The second part of this chapter introduces the fundamental *no-arbitrage principle* saying there should be no free lunch in efficient capital markets. This essential asset pricing argument can be employed in continuous time using the partial-differential-equation(PDE)- and the equivalent-martingale-method(EMM)-approach which we elaborate on two classical cases of asset pricing. Thereafter, we are equipped with the financial modeling framework for building and implementing financial pricing models for valuation and hedging purposes.

1.1 Continuous-Time Stochastics

1.1.1 Stochastic Processes and Brownian Motion

To appropriately set up stochastic differential equations we first outline the conventions of stochastic processes and stochastic calculus in general, defined on a probability space which represents the uncertainty of the intertemporal stochastic economy.

Definition 1.1 (Probability Space) *The underlying set-up for stochastic processes consists of a complete probability space* $(\Omega, \mathcal{F}, \mathbb{P})$*, equipped with a* *filtration, i.e. a non-decreasing family* $\mathbb{F} = \{\mathcal{F}_t\}_{t\in\mathcal{T}}$ *of sub-*σ*-fields of* $\mathcal{F}$: $\mathcal{F}_s \subseteq \mathcal{F}_t \subseteq \mathcal{F}$ *for* $0 \leq s < t < \infty$ *where* $\mathcal{T} = [0,\infty)$. *Here,* $\mathcal{F}_t$ *represents the information available at time t, and the filtration* $\{\mathcal{F}_t\}_{t\in\mathcal{T}}$ *represents the information flow evolving over time and accruing to all agents in the economy.*

Definition 1.2 (Stochastic Process) *A scalar (n-vector) stochastic process, denoted by* $\{\mathbf{X}_t\}_{t\in\mathcal{T}}$, *is a family of random variables (n-vectors) indexed by the parameter set* $\mathcal{T}$, *where the parameter t will refer to time in our applications. The process is defined on a filtered probability space* $(\Omega, \mathcal{F}, \mathbb{P}, \mathbb{F})$ *with values in* $\mathbb{R}^n$. *We say the process is adapted if* $\mathbf{X}_t \in \mathcal{F}_t$ *for each time t, i.e.* $\mathbf{X}_t$ *is* $\mathcal{F}_t$*-measurable; thus,* $\mathbf{X}_t$ *is known when* $\mathcal{F}_t$ *is known. Further, if a filtration is generated by a stochastic process, i.e.* $\mathcal{F}_t = \sigma(\mathbf{X}_s; 0 \leq s \leq t)$, *we call* $\mathcal{F}_t$ *the natural filtration of the process* $\{\mathbf{X}_t\}_{t\in\mathcal{T}}$. *Thus, a process is always adapted to its natural filtration.*

A straightforward classification of stochastic processes is the following: If the random variables (vectors) $\mathbf{X}_t$ are discrete, we say that the stochastic process has a discrete state space. If they are continuous, the process is said to have a continuous state space. The parameter set may also be discrete (for example, $\mathcal{T} = \{1, 2, , \ldots, n\}$, or $\mathcal{T} = \{1, 2, 3, \ldots\}$) or continuous (for example, $\mathcal{T} = [0, 1]$, or $\mathcal{T} = [0, \infty)$). If the parameter set $\mathcal{T}$ is discrete, the stochastic process is a discrete parameter process; however, if it is continuous, we say that the stochastic process is a continuous parameter process. An overview of this classification is shown in table 1.1 following Jazwinski (1970).

Table 1.1. Classification of Stochastic Processes

Classifying Characteristics		Parameter Set $\mathcal{T}$: Discrete	Parameter Set $\mathcal{T}$: Continuous
State Space X	Discrete	Discrete Parameter Chain	Continuous Parameter Chain
State Space X	Continuous	Random Sequence	Stochastic Process

Important to mention is that a stochastic process is measurable with respect to two distinct spaces: For every $t \in \mathcal{T}$ the process $\mathbf{X}_t$ is a random variable with $\omega \to \mathbf{X}_t(\omega); \omega \in \Omega$; if we fix the value for $\omega \in \Omega$ we obtain the mapping $t \to \mathbf{X}_t(\omega); t \in \mathcal{T}$, a so-called path of $\mathbf{X}_t$. Thereby, we need to look at a stochastic process as a function of two distinct variables $(t, \omega) \to \mathbf{X}(t, \omega)$ from $\mathcal{T} \times \Omega$ into the space $I\!R^1$.

Among the most popular building blocks of stochastic models in financial economics is the *standard Brownian motion*, which also underlies the continuous-time pricing models presented in this book.[1] This elementary stochastic process satisfies some distinct properties defined as follows.

Definition 1.3 (Standard Brownian Motion) *A continuous parameter process $\{W_t\}_{t\geq 0}$ is a Brownian motion under the probability measure $\mathbb{P}$ if*

(i) every increment $W_{t+\tau} - W_t$ is normally distributed with $N(\mu\tau, \sigma^2\tau)$ under $\mathbb{P}$, and

(ii) for every pair of disjoint time intervals $[t_1, t_2]$, $[t_3, t_4]$ (with $t_1 < t_2 \leq t_3 < t_4$) the increments $W_2 - W_1$ and $W_4 - W_3$ are independent random variables, and

(iii) $\mathbb{P}(W_0 = 0) = 1$ and W_t is continuous at $t = 0$.

For $\{W_t\}_{t\geq 0}$ to be a standard Brownian motion (or Wiener process), we need to fix the parameters at $\mu = 0$ and $\sigma = 1$.

This means that the increment $W_{t+\tau} - W_t$ is independent of the past, or alternatively, if we know W_t, then further knowledge of the values of W_s for $s < t$ has no effect on our knowledge of the probability law governing $W_{t+\tau} - W_t$. This *Markovian* property of the standard Brownian motion describes a stochastic process for which everything that we know about its future is summarized by its current value.

Definition 1.4 (Markov Process) *A continuous-time stochastic process $\{W_t\}_{t\geq 0}$ is Markovian, if*

$$\mathbb{P}\left[W_t \leq w | W_u, 0 \leq u \leq s\right] = \mathbb{P}\left[W_t \leq w | W_s\right],$$

for all $s < t$.

Furthermore, the standard Brownian motion is a *Gaussian* process, since it has a normal distribution with specific first two moments - the mean $\mu = 0$ and the variance $\sigma^2 = \tau$.[2]

[1] The earliest modeling is by Bachelier (1900), who used this type of stochastic processes to describe stock price movements. To cover jumps in the dynamics of a security, one can add a Poisson process component; see, for example, Merton (1976).

[2] See, for example, Karlin and Taylor (1975, p. 376) and also for further standard distributions Abramowitz and Stegun (1965) and Stuart and Ord (1994).

1.1.2 Martingales, Itô Calculus, and Changes of Measure

In order to quite comfortably solve financial problems in asset pricing we need to make some more formidable stochastic theory accessible after this first conventional outline. The further concepts we look at are: A more general class of processes that can be described as sum of standard mathematical integrals and stochastic integrals - the Itô processes, the martingale property of stochastic processes, the famous Itô formula to solve stochastic integrals and derive asset dynamics dependent on the stochastic behavior of other assets, and finally the important Girsanov theorem to conveniently change probability measures.

Definition 1.5 (Itô Process) *With a standard Brownian motion W_t on $(\Omega, \mathcal{F}, \mathbb{P})$ an Itô process (or stochastic integral) is a stochastic process of the form*

$$X_t = X_0 + \int_0^t \mu(\zeta)\, d\zeta + \int_0^t \sigma(\zeta)\, dW_\zeta, \qquad (1.1)$$

with the drift component μ and the volatility component σ; for both components the integrals corresponding to their second-order moment must be finite.[3]

Equivalently, Itô processes can be symbolically written in differential form $dX_t = \mu dt + \sigma dW_t$, where there are known as *stochastic differential equations*. In equation 1.1 the first integral is an ordinary mathematical (Riemann-Stieltjes) integral whereas the second is a stochastic integral, also known as *Itô integral.*[4] Decomposing the Itô process in a bounded variation component - the ordinary integral - and the martingale component - the Itô integral - is often referred to as the *Doob-Meyer decomposition*, whereby a martingale is formally defined as follows.

Definition 1.6 (Martingale) *On a filtered probability space $(\Omega, \mathcal{F}, \mathbb{P}, \mathbb{F})$ a stochastic process M_t is called a martingale, if the following conditions*

(i) M_t is $\mathcal{F}_t$ measurable for all t,

(ii) $\mathbb{E}[|M_t|] < \infty$ for all t, and

[3] For a complete definition see, for example, Oksendal (2000) who also states the important existence and uniqueness theorem for stochastic differential equations.

[4] For the differences between the two most useful definitions of a stochastic integral - the Itô and the Stratonovich integral - see, for example, Oksendal (2000). There you additionally find the Itô process defined as the limit of a sequence of elementary functions.

(iii) $\mathbb{E}[M_u | \mathcal{F}_t] = M_t$ *for all* $u \geq t$
hold. A process satisfying, for all t *and* u *with* $t \leq u$, *the inequality* $\mathbb{E}[M_u | \mathcal{F}_t] \leq M_t$ ($\mathbb{E}[M_u | \mathcal{F}_t] \geq M_t$) *is called a supermartingale (submartingale).*

We can think of a martingale as a *fair game.* Let us interpret M_n as a gamblers fortune in the discrete case, then the gamblers expected fortune after the (n+1)-th gamble equals the fortune after the n-th gamble, regardless of what may have occurred before. Next, typical financial asset pricing problems in continuous time model asset prices as functions of more than one state variable. Itô's formula, the fundamental instrument in stochastic calculus, allows such functions to be differentiated.

Lemma 1.7 (Itô's Formula) *Let* $\mathbf{X}_t$ *be an* n*-dimensional Itô process in matrix notation of the form*

$$d\mathbf{X}_t = \boldsymbol{\mu} dt + \boldsymbol{\sigma} d\mathbf{W}_t$$

with $\mathbf{X}_t = [X_1, \ldots, X_n]'$, $\quad d\mathbf{W}_t = [dW_{1,t}, \ldots, dW_{m,t}]'$,

$$\boldsymbol{\mu} = \begin{bmatrix} \mu_1 \\ \vdots \\ \mu_n \end{bmatrix}, \qquad \textit{and } \boldsymbol{\sigma} = \begin{bmatrix} \sigma_{11} & \cdots & \sigma_{1m} \\ \vdots & & \vdots \\ \sigma_{n1} & \cdots & \sigma_{nm} \end{bmatrix}$$

and let $\mathbf{g}(t, \mathbf{X}_t) = (g_1(t, \mathbf{X}_t), \ldots, g_p(t, \mathbf{X}_t))$ *be a mapping* $[0, \infty) \times \mathbb{R}^n \to \mathbb{R}^p$. *Then the process*

$$\mathbf{Y}_t = \mathbf{g}(t, \mathbf{X}_t)$$

is again an Itô process whose component Y_k *is given by*

$$dY_k = \frac{\partial g_k(t, \mathbf{X}_t)}{\partial t} dt + \sum_i \frac{\partial g_k(t, \mathbf{X}_t)}{\partial X_i} dX_{i,t} + \frac{1}{2} \sum_{i,j} \frac{\partial^2 g_k(t, \mathbf{X}_t)}{\partial X_i \partial X_j} dX_{i,t} dX_{j,t},$$

where the Itô multiplication rules $dW_{i,t}dt = dtdW_{i,t} = 0$ *and* $dW_{i,t}dW_{j,t} = \rho_{ij}dt$ *apply.*

Thus, stochastic differential calculus is an extension of standard calculus in which we use the *Itô multiplication rules* to account for the effect of nontrivial quadratic variation. Beginning with applying Itô's formula in section 1.2, we will notice its fundamental necessity in financial modeling.

Finally, with the Girsanov theorem we provide the vitally important technique of changing probability measures when, for example, using the EMM-approach in asset pricing. In this connection we need to look at the basic idea of a Radon-Nikodým derivative and equivalent measures.

Definition 1.8 (Radon-Nikodým and Equivalent Measure) *Suppose* ζ *is a non-negative random variable on* $(\Omega, \mathcal{F}, \mathbb{P})$ *with* $\mathbb{E}^{\mathbb{Q}}[\zeta] = 1$, *then we can create a new probability measure* $\mathbb{Q}$

$$\mathbb{Q}(A) = \int\limits_A \zeta d\mathbb{P} = \mathbb{E}^{\mathbb{Q}}[\zeta \mathbf{1}_A]$$

for any event A *with* $\mathbb{Q}$ *being absolutely continuous with respect to* $\mathbb{P}$. *The measurable function* $\frac{d\mathbb{Q}}{d\mathbb{P}} = \zeta$ *is called the Radon-Nikodým derivative of* $\mathbb{Q}$ *with respect to* $\mathbb{P}$. *For a random variable* X *it follows* $\mathbb{E}^{\mathbb{Q}}[X] = \mathbb{E}^{\mathbb{P}}[\zeta X]$. *If* $\mathbb{Q}(A) > 0$ *whenever* $\mathbb{P}(A) > 0$, *and vice versa, then* $\mathbb{P}$ *and* $\mathbb{Q}$ *are said to be equivalent measures, i.e.* $\mathbb{P} \sim \mathbb{Q}$, *they have the same null sets.*

Theorem 1.9 (Girsanov). *Consider the filtered probability space* $(\Omega, \mathcal{F}, \mathbb{P}, \mathbb{F})$ *where* $\{\lambda_t\}_{0 \le t \le T}$ *is an adapted process satisfying Novikov's condition* $\mathbb{E}^{\mathbb{P}}\left[\exp\left(\frac{1}{2}\int_0^T \lambda^2(\zeta)\, d\zeta\right)\right] < \infty$ *such that the following process* $\{M_t\}_{0 \le t \le T}$ *is a* $\mathbb{P}-$*martingale*

$$M_t = \exp\left(-\int\limits_0^t \lambda(\zeta)\, dW_\zeta^{\mathbb{P}} - \frac{1}{2}\int\limits_0^t \lambda^2(\zeta)\, d\zeta\right),$$

with $dW_t^{\mathbb{P}}$ *a standard Brownian motion under the measure* $\mathbb{P}$. *Then under the equivalent probability measure* $\mathbb{Q}$ *with Radon-Nikodým derivative* $\frac{d\mathbb{Q}}{d\mathbb{P}} = M_T$ *the process defined by*

$$W_t^{\mathbb{Q}} = W_t^{\mathbb{P}} + \int\limits_0^t \lambda(\zeta)\, d\zeta$$

is a standard Brownian motion under measure $\mathbb{Q}$, $0 \le t \le T$.[5]

The Girsanov theorem is essential in asset pricing since it allows for a *change in drift* of Itô processes. For example, let $\{X_t\}_{0 \le t \le T}$ be an Itô process of the form

$$dX(t) = \mu dt + dW_t^{\mathbb{P}}$$

and assume $\lambda(t)$ being a constant $\lambda = \mu$, then the change of measure by introducing the Radon-Nikodým derivative $\exp\left(-\mu dW_t^{\mathbb{P}} - \frac{1}{2}\mu^2 dt\right)$ corresponds to

[5] See the original work by Girsanov (1960) and for different versions of the formula, for example, Oksendal (2000).

$$dX(t) = dW_t^{\mathbb{Q}},$$

i.e. $\{X_t\}_{0 \le t \le T}$ is a driftless Brownian motion under the equivalent measure $\mathbb{Q}$. Furthermore, since the process resulted in a martingale under measure $\mathbb{Q}$ it is called an *equivalent martingale measure.*

1.2 Arbitrage Pricing in Continuous Time

A riskless arbitrage opportunity is one that generates non-negative returns under all circumstances and positive returns under some circumstances without any initial investment. In efficient capital markets such opportunities should not exist. This *no-arbitrage principle* is a fundamental pricing concept underlying if not unifying modern asset pricing theory.[6] Under idealized market conditions - such as neglecting transaction costs, liquidity constraints, trading restrictions, and tax issues - asset pricing models need to deliver fair asset values in the absence of arbitrage, i.e. there should be no free lunches in efficient capital markets. The idea behind no-arbitrage pricing is that every modeled security should be hedgeable with a dynamic replicating portfolio. There, we distinguish between dynamically complete and dynamically incomplete pricing models for which asset pricing theory provides the concept of market prices of risk.[7] In the following, we comprehensively touch on arbitrage pricing theory concentrating on the two canonical methods, the PDE- and EMM-approach. Targeted at our modeled applications in parts II, III, and IV we choose to work with two classical asset pricing problems: We examine the *PDE-approach* in an incomplete market situation with deriving a pricing formula for bonds à la Vasicek (1977) where we also touch the basic pricing ideas behind term structure modeling. For the *EMM-approach* we model the arbitrage-free price of an equity option within a complete market à la Black and Scholes (1973) and show in detail how to apply the stochastic calculus made available so far.

1.2.1 PDE Approach

The first major method we consider in calculating an arbitrage-free security price is the *partial-differential-equation*(PDE)-approach. It starts with

---

[6] See, for example, Black and Scholes (1973), Duffie (2001), Dybvig and Ross (1987), Ingersoll (1987), and Varian (1987).

[7] In case of further interest in a detailed analysis of the differences between arbitrage- and equilibrium-oriented pricing see, for example, Schöbel (1995b).

the predefined stochastic evolution of the relevant state variables over time.[8] Thereupon, one can derive the stochastic price dynamics of a generic fixed-income security $P(t,T)$ by applying Itô's formula on the specified state variables. The next step is to implement a self-financing trading strategy using all available capital market securities that are contained in the assumed financial model economy. With no exogenous infusion or withdrawal of money the dynamically rebalanced portfolio should not earn any profit. From a hedging viewpoint a concrete interpretation of this arbitrage pricing is that the arbitrage-free value of a security equals its replicating cost. By ensuring that there are no arbitrage possibilities we come up with what is known as the *fundamental partial differential equation* in finance, which gives the approach its name. Finally, we solve this equation for the security price $P(t,T)$ which now needs to be specified in detail to impose the relevant boundary conditions. For e.g. valuing a pure discount bond we implement $P(T,T) = 1$ in order to model the 'pull-to-par' phenomenon at maturity.

With the objective to elaborate on the PDE approach we pursue the explicit example of calculating an arbitrage-free price in the following. To point at the peculiarities of modeling and pricing assets in an incomplete market such as the market for fixed-income securities, which we deal with in part III, we choose to value a pure discount bond as the basic security from the fixed-income market. We begin with the assumption of the relevant model for the bond market.

Assumption 1.1 (Bond Market) *For the state variable in our model we assume the a priori specified belief of the short-rate dynamics*

$$dr(t) = \mu(r,t)\,dt + \sigma(r,t)\,dW_t^{\mathbb{P}} \tag{1.2}$$

under the objective probability measure $\mathbb{P}$. By assumption the short rate is the only object given a priori, i.e. there exists only one exogenously given asset which is the risk-free asset. We further assume the existence of a market for discount bonds $P(t,T)$ for every choice of maturity T. The bond price is modeled by a function of the short rate $r(t)$ and time t, i.e. $P(t,T) = f(t,r;T)$, where the maturity date T is treated as a constant parameter.

In order to find the arbitrage-free price of the discount bonds, we use the idea of an investor seeking for arbitrage opportunities. He has the possibility

[8] For a comparative overview of different hypothesized distributions in financial modeling see, for example, McDonald (1996); among the first noticing empirical inconsistencies with normally distributed stock returns were Fama (1965) and Mandelbrot (1963).

to hold three different assets: The money market account and two discount bonds $P(t,T_i)$ with any chosen two different maturities $i \in [1,2]$. The money market account, also known as accumulator or savings account, is defined as follows.

Definition 1.2 (Money Market) *For the money market, we define the money account by*

$$B(0,t) = \exp \int_0^t r(\zeta)\, d\zeta,$$

i.e. the account starts with an initial value of $B(0,0) = 1$ and accumulates interest with the dynamics

$$dB(0,t) = r(t)\, B(0,t)\, dt$$

over time.

Using Itô's formula on the above assumption 1.1 of the bond market the discount bonds $P(t,T_1)$ and $P(t,T_2)$ are governed by the price dynamics

$$dP(t,T_i) = \mu_i P(t,T_i)\, dt + \eta_i P(t,T_i)\, dW. \tag{1.3}$$

$$\text{with } \mu_i = \frac{1}{P(t,T_i)} \left[\frac{\partial P(t,T_i)}{\partial t} + \mu \frac{\partial P(t,T_i)}{\partial r} + \frac{1}{2}\sigma^2 \frac{\partial^2 P(t,T_i)}{\partial r^2} \right], \text{ and}$$
$$\eta_i = \frac{1}{P(t,T_i)} \left[\sigma \frac{\partial P(t,T_i)}{\partial r} \right]$$

The portfolio structure of the arbitraguer is given by the sum of the investments in the three tradeable assets in the market:

$$V(t) = n_0(t)\, B(t) + n_1(t)\, P(t,T_1) + n_2(t)\, P(t,T_2),$$

with the absolute portions denoted by n_j with $j \in [1,2,3]$. His portfolio increases or decreases in wealth over the interval $[t, t+dt]$ by

$$\begin{aligned} dV(t) = {} & n_0(t)\, dB(t) + n_1(t)\, dP(t,T_1) + n_2(t)\, dP(t,T_2) \\ & + dn_0(t)\, B(t) + dn_1(t)\, P(t,T_1) + dn_2(t)\, P(t,T_2) \\ & + dn_0(t)\, dB(t) + dn_1(t)\, dP(t,T_1) + dn_2(t)\, dP(t,T_2). \end{aligned} \tag{1.4}$$

To ensure that the formed *portfolio is self-financing* over time, it needs to qualify the condition

$$dn_0(t)(B(t)+dB(t))+dn_1(t)(P(t,T_1)+dP(t,T_1))$$
$$+dn_2(t)(P(t,T_2)+dP(t,T_2))=0.$$

This condition states that the portfolio need to be restructured without requiring any cash inflow and does not permit any outflows which reduces equation (1.4) to:

$$\begin{aligned}dV(t) &= [n_0(t)\,r(t)\,B(0,t)+n_1(t)\,\mu_1 P(t,T_1)+n_2(t)\,\mu_2 P(t,T_2)]\,dt\\ &\quad+[n_1(t)\,\eta_1 P(t,T_1)+n_2(t)\,\eta_2 P(t,T_2)]\,dW.\end{aligned}$$

To constrain this wealth dynamics to an arbitrage-free portfolio we use the following conditions.

Condition 1.3 (No Arbitrage) *Conditions for there being no possibilities of arbitrage:*

(i) Zero investment, i.e. the portfolio is self-financing at the beginning

$$V(0)=n_0 B(0,t)+n_1 P(0,T_1)+n_2 P(0,T_2)=0.$$

(ii) No market risk which requires

$$n_1(t)\,\eta_1 P(t,T_1)+n_2(t)\,\eta_2 P(t,T_2)=0.$$

(iii) This in return means that there can be no return:

$$n_0(t)\,r(t)\,B(0,t)+n_1(t)\,\mu_1 P(t,T_1)+n_2(t)\,\mu_2 P(t,T_2)=0,$$

i.e. the arbitraguer should not be able to realize any profits on his portfolio.

Using the abbreviations $w_0(t)=n_0(t)\,B(0,t)$ and $w_i(t)=n_i(t)\,P(t,T_i)$ for the relative portfolio weights the conditions of no-arbitrage reduce to

$$\begin{bmatrix} 1 & 1 & 1 \\ 0 & \eta_1 & \eta_2 \\ r(t) & \mu_1 & \mu_2 \end{bmatrix}\begin{bmatrix} w_0(t) \\ w_1(t) \\ w_2(t) \end{bmatrix}=0. \tag{1.5}$$

Therein, solutions for the portfolio weights only exist if the market satisfies the internal consistency relation, i.e. the matrix in equation (1.5) needs to be singular. With one degree of freedom we choose, for example, the last row being a linear combination of the remaining rows. Using λ for the free coefficient not determined in the model for the bond market yields the criteria

$$\mu_1=r+\lambda\eta_1,\ \text{and}\ \mu_2=r+\lambda\eta_2.$$

known as the *standard finance condition* which must be satisfied by the returns μ_i and volatilities η_i in no-arbitrage pricing. The equations reduce to the internal consistency result for the bond market

$$\lambda = \frac{\mu_1 - r}{\eta_1} = \frac{\mu_2 - r}{\eta_2} \tag{1.6}$$

in order to avoid arbitrage possibilities. More specifically, the quantity λ is called the *market price of interest rate risk*, since it gives the extra increase in expected instantaneous rate of return $\mu_i - r$ on a bond per an additional unit of risk η_i. The result of equation (1.6) now states that all bonds in an arbitrage-free market have the same market price of risk regardless of the time to maturity. This result contains even more information on the bond dynamics. We substitute the drift and volatility coefficient of equation (1.3) into one of equations (1.6) to obtain the so-called *fundamental partial differential equation*

$$\frac{1}{2}\sigma^2 \frac{\partial^2 P(t,T_i)}{\partial r^2} + (\mu - \lambda\sigma)\frac{\partial P(t,T_i)}{\partial r} + \frac{\partial P(t,T_i)}{\partial t} = rP(t,T_i), \tag{1.7}$$

also known as the *term structure equation* in fixed-income modeling. This equation governs the price of discount bonds in an arbitrage-free market when we additionally incorporate the appropriate boundary condition $P(T_i, T_i) = 1$ for discount bonds. We also note that the price of a particular bond $P(t, T_i)$ is not uniquely given by the bond market model in that the market price of risk λ remains undetermined in the model, i.e. it constitutes an exogenous parameter. The reason for this fact is that arbitrage pricing is always a case of pricing an asset (in our case $P(t, T_i)$) in terms of some other assets (in our case r). But, with the bond market choosing a value for the risk premium λ to satisfy the internal consistency relation of equation (1.6) arbitrage possibilities are avoided. Thus, in the case of the incomplete market for fixed-income securities the equivalent martingale measure $\mathbb{Q}$ with the Girsanov transformation

$$dW_t^{\mathbb{Q}} = dW_t^{\mathbb{P}} + \lambda dt$$

is set by the market, i.e. the risk premium λ can be obtained from observed fixed-income market data.

The further explicit calculation of an arbitrage-free price of discount bonds is based on the one-factor term structure model of Vasicek (1977). There, the short-rate dynamics are defined via the Itô process

$$dr(t) = \kappa(\mu - r)\,dt + \sigma dW_t^{\mathbb{P}},$$

i.e. the short-rate follows a type of Ornstein-Uhlenbeck process. This Gaussian term structure model falls within the class of affine models studied in Duffie and Kan (1996) with known solution

$$P(t,T,r) = e^{A(t)r+B(t)}$$

to the fundamental partial differential equation (1.7). The boundary condition of the discount bond then translates to $A(T) = B(T) = 0$. For the required partial derivatives we obtain

$$P_r = \tfrac{\partial P(t,T,r)}{\partial r} = AP, \qquad P_{rr} = \tfrac{\partial^2 P(t,T,r)}{\partial r^2} = A^2 P, \text{ and}$$

$$P_t = \tfrac{\partial P(t,T,r)}{\partial t} = (A_t r + B_t) P,$$

which we insert into equation (1.7) to yield

$$\tfrac{1}{2}\sigma^2 A^2 P + (\kappa(\mu - r) - \lambda\sigma) AP + (A_t r + B_t) P = rP, \text{ or}$$

$$[A_t - \kappa A - 1] \cdot r + \left[\tfrac{1}{2}\sigma^2 A^2 + (\kappa\mu - \lambda\sigma) A + B_t\right] = 0$$

For this equation to hold for every value of $r(t)$ the expressions in brackets need to be zero, leading us to two ordinary differential equations in $A(t)$ and $B(t)$:

$$A_t - \kappa A - 1 = 0, \text{ and}$$
$$\frac{1}{2}\sigma^2 A^2 + (\kappa\mu - \lambda\sigma) A + B_t = 0.$$

We use the first solution being $A(t) = -\frac{1}{\kappa}\left(1 - e^{-\kappa(T-t)}\right)$ to solve the integral

$$\int_t^T B_\zeta d\zeta = B(T) - B(t)$$

in conjunction with the boundary condition $B(T) = 0$ yielding:

$$B(t) = -\left[\mu - \frac{1}{\kappa}\lambda\sigma - \frac{\sigma^2}{2\kappa^2}\right]((T-t) + A(t)) - \frac{\sigma^2}{4\kappa}A^2(t),$$

which corresponds to the solution of the term structure equation stated in equation (27) of Vasicek (1977). With the derived solutions for $A(t)$ and $B(t)$ based on the appropriate boundary conditions for valuing discount bonds, we finish our elaborated example on the PDE approach. We recognize that the price for discount bonds $P(t, T, r)$ has the market price of risk parameter λ in its coefficient $B(t)$. Assuming this pricing model reasonably fits the observable prices on the bond market, the next step is to implicitly obtain the parameter information from traded market prices using empirical methods. This results in parameter estimates and with the gathered information on the market price of risk also the used martingale measure, what is called model *calibration to market data.*[9]

The PDE pricing approach has been shown to be appealing in two ways: First, it only requires a moderate level of stochastic calculus including applying Itô's formula right at the beginning to specify the assumed economy with the relevant state variables. Second, the main task is to solve the relevant PDE with well-defined corresponding boundary conditions for the particular security. In general, *closed-form solutions* to the PDE are generally preferred, for example, for a comparative statistical analysis in understanding how the value of a security changes relative to changes in a given parameter for hedging purposes. Early examples for analytical asset pricing formulae are Black and Scholes (1973) for the equity market and the analyzed solution for the Vasicek (1977) model for the fixed-income market. In search of closed-form solutions Kamke (1983) and Kamke (1979) provide an extensive and useful collection of different PDEs with known solutions. Otherwise, numerical solutions are always implementable using *finite difference methods* where various explicit and implicit, for example the Crank-Nicholson approximation, schemes are available. These finite difference solutions were introduced in finance to problems of asset pricing by Schwartz (1977), followed by applications of Brennan and Schwartz (1978), Courtadon (1982), and Hull and White (1990b). For computational and implementational aspects of the finite difference methods see, for example, Wilmott, Dewynne, and Howison (1993) and Press, Teukolsky, Vetterling, and Flannery (2002).

1.2.2 EMM Approach

This section continues to investigate the arbitrage pricing theory with elaborating on the *equivalent-martingale-measure-*(EMM)-approach. This method is synonymously called the risk-neutral pricing approach and has proven to be convenient and suitable even for complex valuation problems.[10] Since the

[9] In chapter 2 we examine the estimation principles underlying our applications in parts II, III, and IV.

[10] For a broad variety of applications in equity, fixed-income, and foreign-exchange markets see, for example, Musiela and Rutkowski (1997).

key explanatory works of Harrison and Kreps (1979) and Harrison and Pliska (1981) we know that the no-arbitrage assumption[11] is equivalent to the existence of a risk-neutral probability measure. Thus, the existence of an EMM guarantees the absence of arbitrage opportunities, and the reverse implication also holds so that a model for a securities market is arbitrage-free if and only if there is at least one EMM.

Theorem 1.4 (Fundamental Theorem of Asset Pricing). *The no-arbitrage condition and the existence of an equivalent martingale measure are fundamentally connected as follows:*

(i) the market is arbitrage free, if and only if there exists at least one equivalent martingale measure $\mathbb{Q}$, *and*

(ii) the market is complete, if and only if there exists only one equivalent martingale measure $\mathbb{Q}$

for a modeled capital market.[12]

If a market is free of arbitrage, there exists an equivalent martingale measure on *forward market scenarios* such that the security prices are martingales, after adjusting for the time value of money. This pricing method is an alternative to the PDE-approach, i.e. instead of deriving the value for an asset from a partial differential equation we can equivalently obtain its value by calculating an expectation. Actually there is a link between the expected values of stochastic processes and partial differential equations due to *Feynman-Kac*, who derived the stochastic representation of the solutions to partial differential equations.[13] Formally, the basic proposal of arbitrage pricing under the EMM-approach is the following.

Proposition 1.5 (Risk-Neutral Pricing) *If the market is arbitrage free, there exists a martingale measure* $\mathbb{Q}$ *under which the value of a derivative asset, denoted by the pricing function* $\Pi(t, T, \mathbf{x})$, *is represented by the expectation*

$$\Pi\left(t,T,\mathbf{x}\right)=\mathbb{E}^{\mathbb{Q}}\left[e^{-\int_t^T r(\zeta)d\zeta}\,\Pi\left(T,T,\mathbf{x}\right)\middle|\,\mathcal{F}_t\right], \tag{1.8}$$

[11] I.e. the security payoff is attainable via a replicating strategy whose terminal value equals that of a self-financing portfolio; see section 1.2.1.

[12] The theory is essentially due to Harrison and Pliska (1981), building on the work in discrete time by Harrison and Kreps (1979); see also Stricker (1984) and Delbaen and Schachermayer (1994).

[13] Named after the contributions of Feynman (1948) and Kac (1949). See theorem 15.1.

where the characteristic boundary condition is given by the time T value of the pricing function, i.e. by $\Pi(T,T,\mathbf{x})$.[14]

The peculiarity about the calculation of this expectation is that it is done not with respect to the original measure $\mathbb{P}$ but with respect to the equivalent martingale measure $\mathbb{Q}$. The relevant martingale measure is found by applying a *numéraire asset* which makes the contingent claim a martingale. In evaluating the prices of equity options one conveniently works with the money market account $B(t,T)$ as numéraire.[15] For the further examination of the EMM-approach we choose to work within a Black and Scholes (1973) world which is defined as follows.

Assumption 1.6 (Stock Market) *In the Black and Scholes (1973) model the frictionless economy is given by the a priori specified belief of the stock price dynamics*

$$dS_t = \mu S_t dt + \sigma S_t dW_t^{\mathbb{P}}$$

under the objective probability measure $\mathbb{P}$. By assumption the short rate is the only object given a priori, i.e. there exists only one exogenously given asset which is the risk-free asset $dB_t = rB_t dt$, with $B_0 = 1$.

With this model assumptions the discounted stock price process[16] is given by

$$S_t^* = S_t / B_t$$

$$\text{with: } S_t = S_0 \exp\left[\left(\mu - \tfrac{1}{2}\sigma^2\right)t + \sigma W_t^{\mathbb{P}}\right]$$
$$B_t = \exp\left[rt\right]$$

where S_t follows a *geometric Brownian motion.*[17] Applying Itô's formula on $F(S_t) = S_t^*$ yields

$$dS_t^* = (\mu - r)\, S_t^* dt + \sigma S_t^* dW_t^{\mathbb{P}}$$

[14] For a complete background of this theorem see, for example, Duffie (2001) and Musiela and Rutkowski (1997).

[15] This results in discounted asset prices being martingales with respect to $\mathbb{Q}$; see Harrison and Pliska (1981) and Geman, Karoui, and Rochet (1995). However, in pricing term structure derivatives the forward martingale measure is used as in section 11.2.

[16] For the choice of the money market account as numéraire see, for example, Artzner (1997).

[17] See, for example, Duffie (2001).

and using the Brownian filtration the Girsanov pair

$$\left.\frac{d\mathbb{Q}}{d\mathbb{P}}\right|_{\mathcal{F}_t} = \exp\left(-\int_0^t \lambda(\zeta)\, dW_\zeta^{\mathbb{P}} - \frac{1}{2}\int_0^t \lambda^2(\zeta)\, d\zeta\right)$$

is a pair of *equivalent martingale measures* $\mathbb{P} \sim \mathbb{Q}$ on $\mathcal{F}_t$. By Girsanov's theorem the change of measure $dW_t^{\mathbb{Q}} = dW_t^{\mathbb{P}} + \lambda(t)\, dt$ yields

$$dS_t^* = S_t^* \left((\mu - r - \sigma\lambda(t))\, dt + \sigma dW_t^{\mathbb{Q}}\right)$$

as the $\mathbb{Q}$-dynamics for S_t^*. Since S_t^* has to be a martingale under $\mathbb{Q}$ the drift component need to fulfill the condition $\mu - r - \sigma\lambda(t) = 0$, i.e. we must choose $\lambda = (\mu - r)/\sigma$ as the unique *market price of risk*. This endogenous derivation of the market price of risk results in a *complete market* situation for the Black and Scholes (1973) world. Finally, the discounted stock price dynamics are $dS_t^* = \sigma S_t^* dW_t^{\mathbb{Q}}$, which is a driftless process; the corresponding risk-neutral stock price process is given by

$$\begin{aligned} dS_t &= rS_t dt + \sigma S_t dW_t^{\mathbb{Q}}, \\ \text{with: } S_t &= S_0 e^{\left(r - \frac{1}{2}\sigma^2\right)t + \sigma W_t^Q} \end{aligned} \tag{1.9}$$

under the equivalent martingale measure. The unique measure $\mathbb{Q}$ is defined for the martingale S_t^* and the *martingale characteristic*

$$E^{\mathbb{Q}}\left[S_u^* \mid \mathcal{F}_t\right] = E^{\mathbb{Q}}\left[\frac{S_u}{B_u} \mid \mathcal{F}_t\right] = \frac{S_t}{B_t} = S_t^*$$

holds for all $u > t$; i.e. the expected future value of the martingale equals the current value of the discounted stock price process. According to proposition 1.5 we are able to price various derivative assets depending on their payoff structure. With the assumed constant and known money market dynamics the discount factor can be placed out of the conditional expectation:

$$F(S_t, t) = e^{-r(T-t)} E^{\mathbb{Q}}\left[F(S_T, T) \mid \mathcal{F}_t\right],$$

whereby the derivative asset fulfills the martingale characteristic

$$\begin{aligned} e^{-rt} F(S_t, t) &= E^{\mathbb{Q}}\left[e^{-rT} F(S_T, T) \mid \mathcal{F}_t\right], \text{ and} \\ F^*(S_t, t) &= E^{\mathbb{Q}}\left[F^*(S_T, T) \mid \mathcal{F}_t\right], \end{aligned}$$

i.e. there is no systematic drift. The specific derivative asset for which we derive the arbitrage free price will be a *European call option* with the payoff characteristic $\max\left(S_T - K, 0\right)$ at maturity T. According to the risk-neutral stock price dynamics of equation (1.9) the distribution of S_T is given by S_t multiplied with the exponent of a normally distributed random variable with $N\left(r - \frac{1}{2}\sigma^2\left(T-t\right), \sigma^2\left(T-t\right)\right)$. However, we conveniently use the random variable X following the distribution $N\left(-\frac{1}{2}\sigma^2\left(T-t\right), \sigma^2\left(T-t\right)\right)$ in order to cancel down the discount factor in front of the expectation. Thus, the price of the call option $e^{-r(T-t)}E^{\mathbb{Q}}\left[\left(S_t e^{x+r(T-t)} - K\right)^+ \mid \mathcal{F}_t\right]$ becomes the integral:

$$F\left(S_t, t\right) = \frac{1}{\sqrt{2\pi\sigma^2(T-t)}} \int_a^{+\infty} \left(S_t e^x - K e^{-r(T-t)}\right) e^{-\frac{1}{2}\left(\frac{x+\frac{1}{2}\sigma^2(T-t)}{\sigma\sqrt{T-t}}\right)^2} dx,$$

where the lower boundary of the integral $a = \ln\left(\frac{K}{S_t}\right) - r\left(T-t\right)$ results from the final option payoff $S_t e^{x+r(T-t)} - K > 0$ at maturity after having solved for X. After substituting the variable $\xi = -\frac{x+\frac{1}{2}\sigma^2(T-t)}{\sigma\sqrt{T-t}}$ (or, respectively, $x = -\sigma\sqrt{T-t}\xi - \frac{1}{2}\sigma^2\left(T-t\right)$), we derive with the transformed upper boundary[18] $b = \left(\ln\left(\frac{K}{S_t}\right) + \left(r - \frac{1}{2}\sigma^2\right)\left(T-t\right)\right)/\sigma\sqrt{T-t}$ the call option value as:

$$F\left(S_t, t\right) = \frac{1}{\sqrt{2\pi}} \int_{-\infty}^{b} \left(S_t e^{-\sigma\sqrt{T-t}\xi - \frac{1}{2}\sigma^2(T-t)} - K e^{-r(T-t)}\right) e^{-\frac{1}{2}\xi^2} d\xi \,.$$

In the first term we combine the exponents to $-\frac{1}{2}\left(\xi + \sigma\sqrt{T-t}\right)^2$ and take out the mean $+\sigma\sqrt{T-t}$ and integrate it into the upper boundary. This results in:

$$F\left(S_t, t\right) = S_t \frac{1}{\sqrt{2\pi}} \int_{-\infty}^{b+\sigma\sqrt{T-t}} e^{-\frac{1}{2}\xi^2} d\xi - K e^{-r(T-t)} \frac{1}{\sqrt{2\pi}} \int_{-\infty}^{b} e^{-\frac{1}{2}\xi^2} d\xi \,.$$

Finally, both integrals or expectations stand for cumulated normal distributions[19] and results in the celebrated option pricing formula of Black and Scholes (1973) in familiar notation:

[18] Since the substitution of the variable ξ comprises a change of sign (we substitute $dx = -\sigma\sqrt{T-t}d\xi$) the integral boundaries need to be interchanged: The lower boundary a becomes the upper boundary $b = -\frac{a+\frac{1}{2}\sigma^2(T-t)}{\sigma\sqrt{T-t}}$ and, respectively, the upper boundary $+\infty$ the lower boundary $-\infty$.

[19] Which gives us $N\left(b + \sigma\sqrt{T-t}\right)$ and $N\left(b\right)$.

$$F(S_t, t) = S_t N(d_1) - Ke^{-r(T-t)} N(d_2)$$

$$\text{with: } d_1 = \frac{\ln \frac{S_t}{K} + \left(r + \frac{1}{2}\sigma^2\right)(T-t)}{\sigma\sqrt{T-t}}, \text{ and}$$
$$d_2 = d_1 - \sigma\sqrt{T-t} \quad .$$

which is a solution in closed-from. Important to note is first, that the final pricing formula neither incorporates the expected stock return μ nor some other parameter to account for the investor's risk preferences.[20] This is due to the available replicating strategy for the attainable claim which can be implemented by any investor and is independent of his risk attitudes; see, for example, Duffie (2001). Second, the only parameter which cannot be observed directly in the market - the volatility σ - can be approximated by historical or implicit estimations.[21] Third, the values resulting from derivative asset pricing that feed into risk management involves heavy use of quantitative models which implicates a non-negligeable amount of model risk (see, for example, Green and Figlewski (1999)).

Thereby, we finish our elaboration on the vitally important EMM-approach. A further in-depth treatment and various applications of the method in a more rigorous mathematical manner can be found in, for example, Musiela and Rutkowski (1997) and Avellaneda and Laurence (2000), which are also recommended for readers who want to step beyond the broad derivative asset pricing material found in Briys, Bellalah, Mai, and de Varenne (1998) and Hull (2003).

[20] For an overview of the 'incomplete' models before Black and Scholes (1973), the various model extensions, and for empirical tests see, for example, Smith (1976).

[21] For an analysis of volatility being a meaningful variation parameter see, for example, Bates (1996).

2

Estimation Principles

This chapter introduces the estimation principles underlying our canonical asset pricing framework. For a general financial pricing model we elaborate on the technique of state space modeling, the relevant filtering algorithms, and finally the estimation of model parameters. These building blocks are empirically employed on capital market data in parts II, III, and IV, where we concentrate on specific adaptions of the presented estimation framework.

2.1 State Space Notation

The term '*state space*' is strongly established in statistics and econometrics, however, it has its origins in the field of engineering, where the early developments of the subject took place since the pathbreaking papers by Kalman (1960) and Kalman and Bucy (1961). The state space form is a very popular and useful way to write a dynamic model for a further analysis with the Kalman filter.[1] It is based on two important sets of system equations: The *measurement equation* which relates the state variables to the variables which can only be observed with measurement noise; the second set of equations, called the *transition* or *process equation*, describes the dynamic evolution of the unobservable variables. The vector of unknown parameters, on which the system matrices and error term specifications of the state space form depend on, will be denoted by ψ.[2]

[1] See, for example, Harvey (1989), Aoki (1990), and Hamilton (1994a). For a comprehensive treatment of the state space approach to time series see Durbin and Koopman (2001).

[2] I.e. we treat both the original parameters and the variances of the measurement errors as part of the vector ψ, as we will see further down in section 2.3 on parameter estimation.

Definition 2.1 (Measurement Equation) *For the functional relationship of the measurable observations* $\mathbf{y}_t$ *with the possibly unobservable state vector* $\boldsymbol{\xi}_t$, *we define the linear form*

$$\underset{g\times 1}{\mathbf{y}_t} = \underset{g\times 1}{\mathbf{a}_t(\boldsymbol{\psi})} + \underset{g\times k}{\mathbf{B}_t(\boldsymbol{\psi})}\,\underset{k\times 1}{\boldsymbol{\xi}_t} + \underset{g\times 1}{\boldsymbol{\varepsilon}_t(\boldsymbol{\psi})}, \tag{2.1}$$

with the parameters $\boldsymbol{\psi}$, *an additive component* $\mathbf{a}_t(\boldsymbol{\psi})$, *a multiplicative matrix* $\mathbf{B}_t(\boldsymbol{\psi})$, *and a noise term* $\boldsymbol{\varepsilon}_t(\boldsymbol{\psi})$. *Further, we assume a normal distribution with*

$$\mathbb{E}[\boldsymbol{\varepsilon}_t] = \mathbf{0}, \text{ and}$$
$$\mathbb{E}[\boldsymbol{\varepsilon}_t\boldsymbol{\varepsilon}_s'] = \begin{cases} \mathbf{H}_t(\boldsymbol{\psi}) & \text{for } s=t \\ \mathbf{0} & \text{otherwise} \end{cases}$$

for the error term $\boldsymbol{\varepsilon}_t(\boldsymbol{\psi})$.

Definition 2.2 (Transition Equation) *For the system, that describes the evolution of the state variables* $\boldsymbol{\xi}_t$ *over time, we assume the linear equation*

$$\underset{k\times 1}{\boldsymbol{\xi}_t} = \underset{k\times 1}{\mathbf{c}_t(\boldsymbol{\psi})} + \underset{k\times k}{\boldsymbol{\Phi}_t(\boldsymbol{\psi})}\underset{k\times 1}{\boldsymbol{\xi}_{t-1}} + \underset{k\times 1}{\boldsymbol{\eta}_t(\boldsymbol{\psi})} \tag{2.2}$$

with the transition matrix $\boldsymbol{\Phi}_t(\boldsymbol{\psi})$, *an additive component* $\mathbf{c}_t(\boldsymbol{\psi})$ *and a Gaussian noise term* $\boldsymbol{\eta}_t(\boldsymbol{\psi})$ *with*

$$\mathbb{E}[\boldsymbol{\eta}_t] = \mathbf{0}, \text{ and}$$
$$\mathbb{E}[\boldsymbol{\eta}_t\boldsymbol{\eta}_s'] = \begin{cases} \mathbf{Q}_t(\boldsymbol{\psi}) & \text{for } s=t \\ \mathbf{0} & \text{otherwise.} \end{cases}$$

The model order is given by the dimension of $\boldsymbol{\xi}_t$, *i.e. it is of* k*-th order. Further, we assume independence between the error terms* $\boldsymbol{\varepsilon}_t(\boldsymbol{\psi})$ *and* $\boldsymbol{\eta}_t(\boldsymbol{\psi})$, *and a normally distributed initial state vector* $\boldsymbol{\xi}_0$, *with* $\mathbb{E}[\boldsymbol{\xi}_0] = \boldsymbol{\xi}_{0|0}$ *and* $\mathbb{C}ov[\boldsymbol{\xi}_0] = \boldsymbol{\Sigma}_{0|0}$, *that is assumed independent of the error terms in the sense that* $\mathbb{E}[\boldsymbol{\xi}_0\boldsymbol{\varepsilon}_t'] = \mathbf{0}$ *and* $\mathbb{E}[\boldsymbol{\xi}_0\boldsymbol{\eta}_t'] = \mathbf{0}$ *for all* t. *This system is covariance-stationary provided that all eigenvalues of* $\boldsymbol{\Phi}(\boldsymbol{\psi})$ *lie inside the unit circle.*[3]

These two equations constitute a *linear first-order Gauss-Markov state space representation* for the dynamic behavior of the observable variables $\mathbf{y}_t$.

[3] See, for example, Hamilton (1994b, ch. 10); in the Gaussian state-space model we have *strict stationarity* (see, for example, Hamilton (1994b, p. 45 f.)), since the normal density is completely specified by its first two moments.

This framework admits a particularly tractable solution for the further analysis of filtering.[4] However, it can be further generalized to allow for time-varying system matrices, non-normal disturbances, correlated disturbances and non-linear dynamics, as will be discussed in particular later.[5] Now, for the derivation of the Kalman filter, we just focus on the general system characterized by equations (2.1) and (2.2).

Before we come to specific applications, we need to address the problem of identification which is due to the generality of the state space form defined in equations (2.2) and (2.1). Identification is of considerable *conceptual* interest, since it is a necessary condition for the properties of the maximum likelihood estimates, described in section 2.3, as well as of *practical* interest when we deal with empirical estimation.

Generally, techniques to infer a model from measured data typically contain two steps. First a family of candidate models is decided upon which we call the *modeling step*. In financial applications this step heavily draws on economic theory and mathematics, especially the field of stochastic theory. In a second step, we look for the particular member of this family that optimally describes the information content revealed by the data. In our applications this step is in fact a *parameter estimation* problem in that we maximize a likelihood function based on the prediction error decomposition.[6] Given this general description, identification is a link between the mathematical model world and the real world of data. As such we need to ensure identification on the level of model building before we proceed with the statistical inference. A model is said to be *not identifiable*, if there are observationally equivalent structures of model parameters ψ that imply the same distribution for the observable random outcomes $\mathbf{y}_t$, i.e. there exist different parameter vectors that lead to the same likelihood $L(\mathbf{y}_T, \mathbf{y}_{T-1}, \dots, \mathbf{y}_1; \psi)$. To be more precise we define the concept of identification following Rothenberg (1971).

Definition 2.3 (Concept of Identification) *With denoting F a parametrized cumulative distribution function, ψ the parameter vector contained in the admissible parameter space Ψ, we call the set $\{F(\mathbf{y}_T, \mathbf{y}_{T-1}, \dots, \mathbf{y}_1|\ \psi)\}_{\psi\in\Psi}$ a model and an element $F(\mathbf{y}_T, \mathbf{y}_{T-1}, \dots, \mathbf{y}_1|\psi)$ of this set a structure. A model is called globally identifiable at particular parameter values ψ^0 if*

$$\psi \neq \psi^0 \Rightarrow F(\mathbf{y}_T, \mathbf{y}_{T-1}, \dots, \mathbf{y}_1|\psi) \neq F\left(\mathbf{y}_T, \mathbf{y}_{T-1}, \dots, \mathbf{y}_1|\psi^0\right),$$

[4] See, for example, Jazwinski (1970, ch. 5).

[5] See, for example, Jazwinski (1970), Tanizaki (1996) and Gourieroux and Monfort (1997).

[6] See section 2.3.

i.e. there exist no other parameter values besides the vector ψ^0 that give rise to the same structure. A model is said to be locally identified at ψ^0 if there exists a $\delta > 0$ such that for any value of ψ satisfying $(\psi - \psi^0)'(\psi - \psi^0) < 0$ there exist realizations $\mathbf{y}_T, \mathbf{y}_{T-1}, \ldots, \mathbf{y}_1$ for which we obtain two different structures.

Dealing with an unidentifiable parameter vector ψ is a severe problem, since alternative parameter structures usually lead to different causal interpretations in the context of the investigated model. Thus the uniqueness of the likelihood must be guaranteed for all possible parameter vectors, because the true parameter structure is not known. For an unidentified model example in the case of maximum likelihood estimation using the Kalman filter see Hamilton (1994b, p. 387 f.).

2.2 Filtering Algorithms

According to the assumed state space form of equations (2.1) and (2.2), we specify stochastic processes of explaining factors $\boldsymbol{\xi}_t$, which describe the state of the system. Instead of being able to observe the factors directly, we can only observe some noisy function $\mathbf{y}_t$ of $\boldsymbol{\xi}_t$. The problem of determining the state of the system from noisy measurements $\mathbf{y}_t$ is called *estimation*. The special estimation problem of *filtering* has the object of obtaining an expression for the optimal estimate of $\boldsymbol{\xi}_t$ given the observations up to time t. The most successful result of this kind is that obtained for linear systems by Kalman (1960), Kalman and Bucy (1961), and Kalman (1963) which we are further dealing with. In general, the estimation problem can be classified upon the available and processed information into three different problems according to the following definition.

Definition 2.1 (Estimation) *Considering the problem of estimating $\boldsymbol{\xi}_t$ using information up to time s, denoted by the information set $\mathcal{F}_s = \{\mathbf{y}_s, \ldots, \mathbf{y}_2, \mathbf{y}_1\}$, we differentiate between the three cases of a*

$$\begin{aligned} \textit{filtering problem}: &\quad \textit{for } t = s, \\ \textit{smoothing problem}: &\quad \textit{for } t < s, \textit{ and} \\ \textit{prediction problem}: &\quad \textit{for } t > s, \end{aligned}$$

dependent on which information set we use in estimation.[7]

[7] See, for example, Jazwinski (1970, p. 142 f.) and Harvey (1993, sec. 4.2).

Furthermore, we need a clear concept of what constitutes a statistically *optimal estimate* in our context. The concept we use is the statistical criterium of mean square error which we define next. We also present the solution to the problem of the *optimal estimator*. Finally, we introduce convenient notations for the exposition of the Kalman filter derivation.

Definition 2.2 (Optimality Criterium) *Let $\boldsymbol{\xi}^*_{t|t-1}$ denote an estimate of $\boldsymbol{\xi}_t$ based on the information set $\mathcal{F}_{t-1}$. In order to choose the optimal of various possible forecasts, we need to specify a criterion of what optimal means. Very convenient results are obtained from assuming a quadratic loss function $L(\mathbf{e}_t)$ on the estimation error $\mathbf{e}_t = \boldsymbol{\xi}_t - \boldsymbol{\xi}^*_{t|t-1}$.[8] Choosing the estimator as to minimize the mean square error*

$$\mathbb{MSE}\left(\boldsymbol{\xi}^*_{t|t-1}\right) \equiv \mathbb{E}\left[\boldsymbol{\xi}_t - \boldsymbol{\xi}^*_{t|t-1}\right]^2$$

results in the minimum mean square estimator as the best or optimal estimator with respect to any quadratic function of the estimation error.

Theorem 2.3 (Optimal Estimator). *Given the availability of an information set $\mathcal{F}_{t-1}$, the optimal estimator $\boldsymbol{\xi}_{t|t-1}$ of $\boldsymbol{\xi}_t$ among all estimators $\boldsymbol{\xi}^*_{t|t-1}$ is the expected value of $\boldsymbol{\xi}_{t-1}$ conditional on the information at time $t-1$; thus the MMSE is given by $\xi_{t|t-1} = \mathbb{E}\left[\boldsymbol{\xi}_t|\mathcal{F}_{t-1}\right]$.[9] The corresponding MSE is given by*

$$\mathbb{MSE}\left(\boldsymbol{\xi}_{t|t-1}\right) = \mathbb{E}\left[\boldsymbol{\xi}_t - \mathbb{E}\left[\boldsymbol{\xi}_t|\mathcal{F}_{t-1}\right]\right]^2 = \mathbb{C}ov\left[\boldsymbol{\xi}_t|\mathcal{F}_{t-1}\right],$$

i.e. in this case the mean square error matrix is equal to the variance-covariance matrix.

Notation 2.4 (Conditional Expectation and Variance) *In considering the problem of estimating $\boldsymbol{\xi}_t$ using information up to time s, i.e. the information set $\mathcal{F}_s = \{\mathbf{y}_s, \ldots, \mathbf{y}_2, \mathbf{y}_1\}$, we denote the conditional expectation of $\boldsymbol{\xi}_t$ given $\mathcal{F}_s$ in the further analysis for convenience by $\mathbb{E}\left[\boldsymbol{\xi}_t|\mathcal{F}_s\right] \equiv \boldsymbol{\xi}_{t|s}$. For the second conditional moment we will further use $\mathbb{C}ov\left[\boldsymbol{\xi}_t|\mathcal{F}_s\right] \equiv \Sigma_{t|s}$, the conditional variance-covariance matrix of $\boldsymbol{\xi}_t$ given $\mathcal{F}_s$.*

2.2.1 Filtering Objective

Having gone through the necessary preliminaries, we will see in this section that the Kalman filter is a *set of equations* which allows an estimator to be

[8] See, for example, Jazwinski (1970, p. 146 f.).
[9] See, for example, Hamilton (1994b, p. 72 f.).

updated once new observational information becomes available. This filtering objective is carried out in two distinct parts:

1. **Prediction Step:** The first step consists of forming an optimal predictor of the next observation, given all the information available up to time $t-1$. We extrapolate the state vector by means of conditional expectation utilizing the information set $\mathcal{F}_{t-1}$ and calculate a so-called *a priori estimate* for time t.
2. **Updating Step:** The a priori state estimate is updated with the new information arriving at time t that is combined with the already available information from time $t-1$. The result of this step is called the *filtered estimate* or the *a posteriori estimate.* Therein the Kalman gain in estimation is realized.

We will find that the Kalman filter provides an *optimal* solution to the presented problems of prediction and updating. In deriving the Kalman filter we can choose among several approaches available in the literature that are linked to different interpretations of the filter. Besides the original derivation in the work by Kalman (1960), where he used the idea of orthogonal projection (see, for example, Aoki (1990) and Brockwell and Davies (1987)), we can derive the Kalman filter from the properties of a multivariate normal distribution as shown, for example, in Harvey (1989), while Burridge and Wallis (1988) and Hamilton (1994b), for example, deduce the Kalman filter to be the minimum mean square linear estimator. Moreover, alternative derivations are found in Jazwinski (1970). In the following we present the elementary derivation of the Kalman filter under the normality assumption which yields the interpretation of the Kalman filter as the optimal filter in the sense of a *minimum mean square estimator.* Also, we obtain the Kalman filter for the general case of non-normality by exploiting the given linear relationships of the observations and dynamics stated respectively in the state space equations (2.1) and (2.2); in this case we speak of the Kalman filter as the optimal filter in the sense of a *minimum mean square linear estimator.*

2.2.2 Optimal Estimator

In order for the Kalman filter to yield a *minimum mean square estimator* (MMSE), we assume that the additive error terms ε_t and η_t are independently and normally distributed. Furthermore, the error terms are treated as independent of the initial state vector, which is assumed to be normally distributed with $\mathbb{E}[\boldsymbol{\xi}_0] = \boldsymbol{\xi}_0$ and $\mathbb{C}ov[\boldsymbol{\xi}_0] = \Sigma_0$. Therefore, since the transition equation (2.2) is linear in $\boldsymbol{\xi}_{t-1}$ and the error term $\boldsymbol{\eta}_t$, the state vector $\boldsymbol{\xi}_t$ - as the sum of $\boldsymbol{\xi}_{t-1}$ and $\boldsymbol{\eta}_t$ - is also normally distributed. Furthermore, for the measurement equation (2.1) we also have a normally distributed $\mathbf{y}_t$, since $\boldsymbol{\xi}_t$ and ε_t are both normal. Thus, we are dealing with a model with Gaussian signals perturbed by additive Gaussian noise.

In the *prediction step*, we forecast the state vector by calculating the conditional expectation of the state variables on both sides of equation (2.2) given the information up to time $t-1$:

$$\boldsymbol{\xi}_{t|t-1} \equiv \mathbb{E}\left[\boldsymbol{\xi}_t|\mathcal{F}_{t-1}\right] = \mathbf{c}_t + \Phi_t \boldsymbol{\xi}_{t-1|t-1}. \tag{2.3}$$

The corresponding variance-covariance matrix for the state variables is given by

$$\begin{aligned}
\Sigma_{t|t-1} &\equiv \mathbb{E}\left[\left(\boldsymbol{\xi}_t - \mathbb{E}\left[\boldsymbol{\xi}_t|\mathcal{F}_{t-1}\right]\right)\left(\boldsymbol{\xi}_t - \mathbb{E}\left[\boldsymbol{\xi}_t|\mathcal{F}_{t-1}\right]\right)'|\mathcal{F}_{t-1}\right] \\
&= \mathbb{E}\left[\left(\boldsymbol{\xi}_t - \boldsymbol{\xi}_{t|t-1}\right)\left(\boldsymbol{\xi}_t - \boldsymbol{\xi}_{t|t-1}\right)'|\mathcal{F}_{t-1}\right] \\
&= \mathbb{E}\Big[\left(\mathbf{c}_t + \Phi_t\boldsymbol{\xi}_{t-1} + \boldsymbol{\eta}_t - \mathbf{c}_t - \Phi_t\boldsymbol{\xi}_{t-1|t-1}\right) \\
&\qquad \left(\mathbf{c}_t + \boldsymbol{\xi}_{t-1}'\Phi_t' + \boldsymbol{\eta}_t - \mathbf{c}_t - \boldsymbol{\xi}_{t-1|t-1}'\Phi_t'\right)|\mathcal{F}_{t-1}\Big] \\
&= \Phi_t\mathbb{E}\left[\boldsymbol{\xi}_{t-1|t-1}\boldsymbol{\xi}_{t-1|t-1}'|\mathcal{F}_{t-1}\right]\Phi_t' + \mathbb{E}\left[\boldsymbol{\eta}_t\boldsymbol{\eta}_t'|\mathcal{F}_{t-1}\right] \\
&= \Phi_t\Sigma_{t-1|t-1}\Phi_t' + \mathbf{Q}_t.
\end{aligned} \tag{2.4}$$

Equations (2.3) and (2.4) are known as the prediction equations of the Kalman filter.

As an intermediate result we define the prediction error denoted by $\mathbf{v}_t$, using the given data for the measurable observations $\mathbf{y}_t$, as

$$\mathbf{v}_t = \mathbf{y}_t - \mathbf{y}_{t|t-1} = \mathbf{y}_t - \mathbb{E}\left[\mathbf{y}_t|\mathcal{F}_{t-1}\right] = \mathbf{y}_t - \mathbf{a}_t - \mathbf{B}_t\boldsymbol{\xi}_{t|t-1},$$

and are able to calculate the variance-covariance matrix of $\mathbf{v}_t$ as

$$\begin{aligned}
\mathbf{F}_{t|t-1} &= \mathbb{C}ov\left[\mathbf{v}_t|\mathcal{F}_{t-1}\right] \\
&= \mathbb{E}\left[\left(\mathbf{y}_t - \mathbb{E}\left[\mathbf{y}_t|\mathcal{F}_{t-1}\right]\right)\left(\mathbf{y}_t - \mathbb{E}\left[\mathbf{y}_t|\mathcal{F}_{t-1}\right]\right)'|\mathcal{F}_{t-1}\right] \\
&= \mathbb{E}\Big[\left(\mathbf{a}_t + \mathbf{B}_t\boldsymbol{\xi}_t + \varepsilon_t - \mathbf{a}_t - \mathbf{B}_t\boldsymbol{\xi}_{t|t-1}\right) \\
&\qquad \left(\mathbf{a}_t' + \boldsymbol{\xi}_t'\mathbf{B}_t' + \varepsilon_t' - \mathbf{a}_t' - \boldsymbol{\xi}_{t|t-1}'\mathbf{B}_t'\right)|\mathcal{F}_{t-1}\Big] \\
&= \mathbf{B}_t\mathbb{E}\left[\boldsymbol{\xi}_{t|t-1}\boldsymbol{\xi}_{t|t-1}'|\mathcal{F}_{t-1}\right]\mathbf{B}_t' + \mathbb{E}\left[\varepsilon_t\varepsilon_t'|\mathcal{F}_{t-1}\right] \\
&= \mathbf{B}_t\Sigma_{t|t-1}\mathbf{B}_t' + \mathbf{H}_t.
\end{aligned} \tag{2.5}$$

In the last step, the *updating step*, we update the inference on $\boldsymbol{\xi}_{t|t-1}$ by including the newly available information at time t, which results in the filtered estimate $\boldsymbol{\xi}_{t|t}$. Hereby, we can realize the so-called Kalman gain $\mathbf{K}_t$. In order to obtain the updating equations, we first consider the joint distribution of $\boldsymbol{\xi}_t$ and $\mathbf{y}_t$. Since both variables are normally distributed, the joint distribution given the information $\mathcal{F}_{t-1}$ is

$$\begin{bmatrix} \boldsymbol{\xi}_t \\ \mathbf{y}_t \end{bmatrix} \sim N\left(\begin{bmatrix} \boldsymbol{\xi}_{t|t-1} \\ \mathbf{y}_{t|t-1} \end{bmatrix}, \begin{bmatrix} \Sigma_{t|t-1} & \Sigma_{t|t-1}\mathbf{B}_t' \\ \mathbf{B}_t\Sigma_{t|t-1} & \mathbf{F}_{t|t-1} \end{bmatrix} \right) \tag{2.6}$$

with the conditional means

$$\begin{aligned} \boldsymbol{\xi}_{t|t-1} &\equiv \mathbb{E}\left[\boldsymbol{\xi}_t|\mathcal{F}_{t-1}\right] = \mathbf{c}_t + \Phi_t\boldsymbol{\xi}_{t-1|t-1} \\ \mathbf{y}_{t|t-1} &\equiv \mathbb{E}\left[\mathbf{y}_t|\mathcal{F}_{t-1}\right] = \mathbf{a}_t + \mathbf{B}_t\boldsymbol{\xi}_{t|t-1}, \end{aligned}$$

and the conditional variance-covariance matrices stated in equations (2.4) and (2.5), and

$$\begin{aligned} \mathbb{E}\left[\left(\boldsymbol{\xi}_t - \boldsymbol{\xi}_{t|t-1}\right)\left(\mathbf{y}_t - \mathbf{y}_{t|t-1}\right)'|\mathcal{F}_{t-1}\right] &= \mathbb{E}\left[\left(\boldsymbol{\xi}_t - \boldsymbol{\xi}_{t|t-1}\right)\right. \\ &\qquad \left.\left(\boldsymbol{\xi}_t - \boldsymbol{\xi}_{t|t-1}\right)'|\mathcal{F}_{t-1}\right]\mathbf{B}_t' = \Sigma_{t|t-1}\mathbf{B}_t' \\ \mathbb{E}\left[\left(\mathbf{y}_t - \mathbf{y}_{t|t-1}\right)\left(\boldsymbol{\xi}_t - \boldsymbol{\xi}_{t|t-1}\right)'|\mathcal{F}_{t-1}\right] &= \mathbf{B_t}\Sigma_{t|t-1}. \end{aligned}$$

For the further derivation, we take advantage of the following result for normal variables.

Lemma 2.5 (Conditional Normal Distribution) [10] *Let $\mathbf{z}_1$ and $\mathbf{z}_2$ denote $(n_1\times 1)$ and $(n_2 \times 1)$ vectors of random variables respectively that follow a joint normal distribution:*

$$\begin{bmatrix} \mathbf{z}_1 \\ \mathbf{z}_2 \end{bmatrix} \sim N\left(\begin{bmatrix} \boldsymbol{\mu}_1 \\ \boldsymbol{\mu}_2 \end{bmatrix}, \begin{bmatrix} \Omega_{11} & \Omega_{12} \\ \Omega_{12}' & \Omega_{22} \end{bmatrix} \right).$$

Then the distribution of $\mathbf{z}_1$ conditional on $\mathbf{z}_2$ is $N(\mathbf{m}, \Sigma)$ where

$$\begin{aligned} \mathbf{m} &= \boldsymbol{\mu}_1 + \Omega_{12}\Omega_{22}^{-1}\left(\mathbf{z}_2 - \boldsymbol{\mu}_2\right), \text{ and} \\ \Sigma &= \Omega_{11} - \Omega_{12}\Omega_{22}^{-1}\Omega_{12}'. \end{aligned}$$

Thus the optimal forecast of $\mathbf{z}_1$ conditional on having observed $\mathbf{z}_2$ is given by

[10] See, for example, Gourieroux and Monfort (1995a, p. 481 f.) and for a proof Anderson and Moore (1979).

$$\mathbb{E}\left[\mathbf{z}_1|\mathbf{z}_2\right] = \mathbf{m}$$

with Σ characterizing the mean square error of this forecast:

$$\mathbb{E}\left[\left(\mathbf{z}_1 - \mathbf{m}\right)\left(\mathbf{z}_1 - \mathbf{m}\right)' |\mathbf{z}_2\right] = \Sigma.$$

Applying this result to equation (2.6) yields the following distribution for $\boldsymbol{\xi}_t$ given the information $\mathcal{F}_t$ of the observable data $\mathbf{y}_t$:

$$\begin{aligned}\boldsymbol{\xi}_t \sim N\Big(&\boldsymbol{\xi}_{t|t-1} + \Sigma_{t|t-1}\mathbf{B}_t'\mathbf{F}_{t|t-1}^{-1}\left(\mathbf{y}_t - \mathbf{y}_{t|t-1}\right),\\ &\Sigma_{t|t-1} - \Sigma_{t|t-1}\mathbf{B}_t'\mathbf{F}_{t|t-1}^{-1}\mathbf{B}_t\Sigma_{t|t-1}\Big).\end{aligned}$$

Comparing this distribution with the result for the conditional moments of $\boldsymbol{\xi}_t$ given the information $\mathcal{F}_t$,

$$\boldsymbol{\xi}_t \sim N\left(\boldsymbol{\xi}_{t|t},\ \Sigma_{t|t}\right),$$

we obtain the following relationships, where v_t denotes the prediction error:

$$\begin{aligned}\boldsymbol{\xi}_{t|t} &= \boldsymbol{\xi}_{t|t-1} + \Sigma_{t|t-1}\mathbf{B}_t'\mathbf{F}_{t|t-1}^{-1}\mathbf{v}_t\\ \Sigma_{t|t} &= \Sigma_{t|t-1} - \Sigma_{t|t-1}\mathbf{B}_t'\mathbf{F}_{t|t-1}^{-1}\mathbf{B}_t\Sigma_{t|t-1}.\end{aligned}$$

This finally results in the update for the state vector

$$\boldsymbol{\xi}_{t|t} = \mathbb{E}\left[\boldsymbol{\xi}_t|\mathcal{F}_t\right] = \boldsymbol{\xi}_{t|t-1} + \mathbf{K}_t\mathbf{v}_t \tag{2.7}$$

and its variance-covariance matrix

$$\begin{aligned}\Sigma_{t|t} &= \mathbb{E}\left[\left(\boldsymbol{\xi}_t - \boldsymbol{\xi}_{t|t}\right)\left(\boldsymbol{\xi}_t - \boldsymbol{\xi}_{t|t}\right)' |\mathcal{F}_t\right]\\ &= \Sigma_{t|t-1} - \mathbf{K}_t\mathbf{B}_t\Sigma_{t|t-1}\\ &= \left(I - \mathbf{K}_t\mathbf{B}_t\right)\Sigma_{t|t-1},\end{aligned} \tag{2.8}$$

where we substituted $\mathbf{K}_t = \Sigma_{t|t-1}\mathbf{B}_t'\mathbf{F}_{t|t-1}^{-1}$ for the Kalman gain matrix.

Dealing with a non-normally distributed state vector but confining ourselves at the same time to the class of linear estimators, we derive the Kalman filter as the *minimum mean square linear estimator* (MMSLE) without requiring the normality assumption for the error terms. The prediction equations (2.3) and (2.4) are obtained similarly to the previous derivation by taking the conditional expectation and variance given the information up to time $t-1$.

For obtaining the updating equations, we begin with stating a general linear relationship of the updated estimator $\boldsymbol{\xi}_{t|t}$ and the information of the present sample $(\mathbf{y}_t, \mathbf{a}_t, \mathbf{c}_t)$ and the past information $\mathcal{F}_{t-1}$. We will assume the general linear form

$$\boldsymbol{\xi}_{t|t} = K_t \mathbf{y}_t + \mathbf{L}\mathbf{a}_t + \mathbf{M}\mathbf{c}_t + \mathbf{N}\boldsymbol{\xi}_{t-1|t-1}$$

with arbitrary matrices $\mathbf{K}_t$, $\mathbf{L}$, $\mathbf{M}$, and $\mathbf{N}$. These matrices are now to be chosen such that $\boldsymbol{\xi}_{t|t}$ is the MMSLE; therefore we define the estimation error

$$\begin{aligned}
\mathbf{e}_t &= \boldsymbol{\xi}_t - \boldsymbol{\xi}_{t|t} \\
&= \mathbf{c}_t + \Phi\boldsymbol{\xi}_{t-1} + \boldsymbol{\eta}_t - \mathbf{K}_t\left(\mathbf{a}_t + \mathbf{B}_t\left(\mathbf{c}_t + \Phi_t\boldsymbol{\xi}_{t-1} + \boldsymbol{\eta}_t\right) + \boldsymbol{\varepsilon}_t\right) \\
&\quad -\mathbf{L}\mathbf{a}_t - \mathbf{M}\mathbf{c}_t - \mathbf{N}\left(\boldsymbol{\xi}_{t-1} - \mathbf{e}_{t-1}\right) \\
&= \left(\Phi_t - \mathbf{N} - \mathbf{K}_t\mathbf{B}_t\Phi\right)\boldsymbol{\xi}_{t-1} - \left(\mathbf{L} + \mathbf{K}_t\right)\mathbf{a}_t + \left(\mathbf{I} - \mathbf{M} - \mathbf{K}_t\mathbf{B}_t\right)\mathbf{c}_t \\
&\quad +\mathbf{N}\mathbf{e}_{t-1} + \left(\mathbf{I} - \mathbf{K}_t\mathbf{B}_t\right)\boldsymbol{\eta}_t - \mathbf{K}_t\boldsymbol{\varepsilon}_t. \qquad (2.9)
\end{aligned}$$

For $\boldsymbol{\xi}_{t|t}$ being the MMSLE the estimation error $\mathbf{e}_t$ needs to be (i) unconditionally unbiased, i.e. the estimation error has zero expectation, which gives us the conditions

$$\begin{aligned}
\Phi_t - \mathbf{N} - \mathbf{K}_t\mathbf{B}_t\Phi &= 0, \\
\mathbf{L} + \mathbf{K}_t &= 0, \text{ and} \\
\mathbf{I} - \mathbf{M} - \mathbf{K}_t\mathbf{B}_t &= 0
\end{aligned}$$

for equation (2.9). This yields, with the substitution $\mathbf{N} = \left(\mathbf{I} - \mathbf{K}_t\mathbf{B}_t\right)\Phi_t$, the estimation error

$$\mathbf{e}_t = \left(\mathbf{I} - \mathbf{K}_t\mathbf{B}_t\right)\left(\Phi_t\mathbf{e}_{t-1} - \boldsymbol{\eta}_t\right) - \mathbf{K}_t\boldsymbol{\varepsilon}_t.$$

Furthermore, the estimation error $\mathbf{e}_t$ is required to have (ii) minimum variance, i.e. we need to minimize the estimation errors variance-covariance matrix

$$\begin{aligned}
\Sigma_{t|t} &= \mathbb{E}\left[\left(\left(\mathbf{I} - \mathbf{K}_t\mathbf{B}_t\right)\left(\Phi_t\mathbf{e}_{t-1} - \boldsymbol{\eta}_t\right) - \mathbf{K}_t\boldsymbol{\varepsilon}_t\right)\right. \\
&\qquad \left.\left(\left(\mathbf{I} - \mathbf{K}_t\mathbf{B}_t\right)\left(\Phi_t\mathbf{e}_{t-1} - \boldsymbol{\eta}_t\right) - \mathbf{K}_t\boldsymbol{\varepsilon}_t\right)'\right] \\
&= \left(\mathbf{I} - \mathbf{K}_t\mathbf{B}_t\right)\left(\Phi_t\Sigma_{t-1|t-1}\Phi_t' + \mathbf{Q}_t\right)\left(\mathbf{I} - \mathbf{K}_t\mathbf{B}_t\right)' + \mathbf{K}_t\mathbf{H}_t\mathbf{K}_t' \\
&= \left(\mathbf{I} - \mathbf{K}_t\mathbf{B}_t\right)\Sigma_{t|t-1}\left(\mathbf{I} - \mathbf{K}_t\mathbf{B}_t\right)' + \mathbf{K}_t\mathbf{H}_t\mathbf{K}_t'
\end{aligned}$$

with respect to $\mathbf{K}_t$; for the partial derivative we get:[11]

[11] For derivatives of functions with arguments in matrix form see, for example, Lütkepohl (1996).

$$\begin{aligned}\frac{\partial \Sigma_{t|t}}{\partial \mathbf{K}_t} &= \frac{\partial}{\partial \mathbf{K}_t}\left(\Sigma_{t|t-1} - \mathbf{K}_t\mathbf{B}_t\Sigma_{t|t-1} - \Sigma_{t|t-1}\mathbf{B}_t'\mathbf{K}_t'\right.\\ &\qquad \left.+\mathbf{K}_t\mathbf{B}_t\Sigma_{t|t-1}\mathbf{B}_t'\mathbf{K}_t' + \mathbf{K}_t\mathbf{H}_t\mathbf{K}_t'\right)\\ &= -\left(\mathbf{B}_t\Sigma_{t|t-1}\right)' - \Sigma_{t|t-1}\mathbf{B}_t' + 2\mathbf{B}_t\Sigma_{t|t-1}\mathbf{B}_t'K_t + 2\mathbf{H}_t\mathbf{K}_t\\ &= -2\Sigma_{t|t-1}\mathbf{B}_t' + 2\mathbf{B}_t\Sigma_{t|t-1}\mathbf{B}_t'\mathbf{K}_t + 2\mathbf{H}_t\mathbf{K}_t.\end{aligned}$$

Setting the result equal to zero according to the necessary minimization condition, we can solve for the Kalman gain matrix

$$\mathbf{K}_t = \Sigma_{t|t-1}\mathbf{B}_t'\left(\mathbf{B}_t\Sigma_{t|t-1}\mathbf{B}_t' + \mathbf{H}_t\right)^{-1}$$

which is equivalent to the expression derived in equation (2.8) for the case of the Kalman filter resulting in the MMSE.

2.2.3 Filter Recursions

Having derived the *Kalman filter algorithm* in two different ways yielding the MMSE and the MMSLE, we are now able to apply the sequence of filtering equations *recursively* as each new observation becomes available. Please compare the flowchart of figure 2.1 for a graphical illustration of the Kalman filter algorithm. The recursion algorithm starts with a feasible choice of the parameter vector ψ and corresponding state vector $\boldsymbol{\xi}_0$ with variance-covariance matrix Σ_0. Thereupon, we use the prediction equations to calculate the a priori estimates

$$\begin{aligned}\boldsymbol{\xi}_{t|t-1} &= \mathbf{c}_t + \Phi_t\boldsymbol{\xi}_{t-1|t-1}, \text{ and}\\ \Sigma_{t|t-1} &= \Phi_t\Sigma_{t-1|t-1}\Phi_t' + \mathbf{Q}_t.\end{aligned}$$

Using the current market information on observable variables $\mathbf{y}_t$ on date $t = 1$, we next derive the prediction error and the corresponding MSE as

$$\begin{aligned}\mathbf{v}_t &= \mathbf{y}_t - \mathbf{a}_t - \mathbf{B}_t\boldsymbol{\xi}_{t|t-1}, \text{ and}\\ \mathbf{F}_{t|t-1} &= \mathbf{B}_t\Sigma_{t|t-1}\mathbf{B}_t' + \mathbf{H}_t.\end{aligned}$$

Given these intermediate results we then update the a priori estimates yielding the filtered estimates

$$\begin{aligned}\boldsymbol{\xi}_{t|t} &= \boldsymbol{\xi}_{t|t-1} + \mathbf{K}_t\mathbf{v}_t\\ \Sigma_{t|t} &= \left(\mathbf{I} - \mathbf{K}_t\mathbf{B}_t\right)\Sigma_{t|t-1}\end{aligned}$$

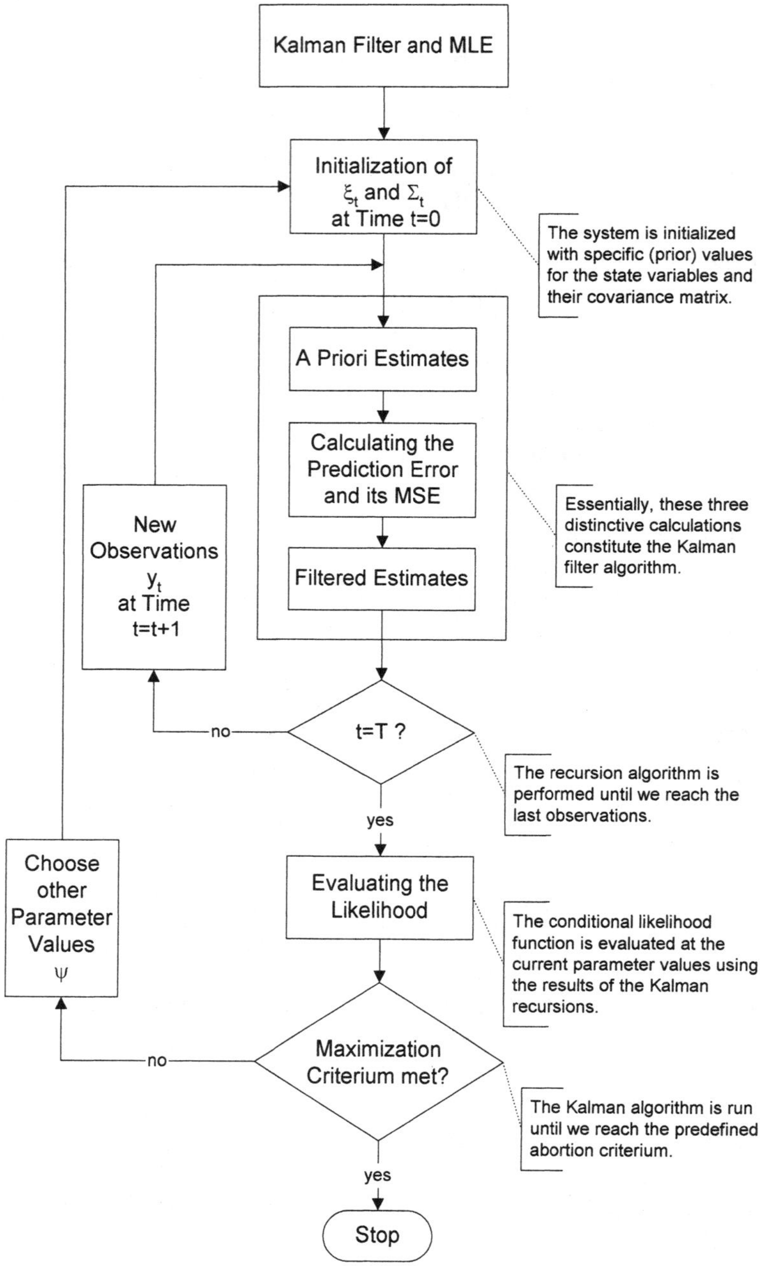

Fig. 2.1. Flowchart of the Kalman Filter and MLE

as optimal in the sense of MMSE or MMSLE using the Kalman gain $\mathbf{K}_t$. Now for each discrete time step, we recursively feed the results for the state vector $\boldsymbol{\xi}_{t|t}$ and its variance-covariance matrix $\Sigma_{t|t}$ from the updating equations into the prediction equations until we reach the last observations $\mathbf{y}_T$.[12] Finally, with the obtained time-series of the state variables we can evaluate the likelihood function as we describe in section 2.3 to choose other more appropriate values for the parameters $\boldsymbol{\psi}$.

2.2.4 Extended Kalman Filtering

In the previous section we have been dealing with the case of linear measurement and transition equations and normally distributed error terms as specified in equations (2.1) and (2.2). However, some applications derived from financial theory exhibit non-linear functional relationships of non-normally distributed state variables in the measurement and transition equations. In such cases explicit expressions for the filtering algorithms cannot be derived, i.e. some approximations are necessary for the estimation procedure. There are two main approaches to obtain non-linear filtering algorithms.[13] The first approach is to approximate the non-linear measurement and transition equations. The linearized non-linear functions are then applied to a modification of the linear Kalman filter algorithm as derived in the previous section. The second types of algorithms can be summarized under the approach of approximating the underlying density functions of the state vector. Then a recursive algorithm on the densities is derived using Bayes's formula. The advantage of this approach is that it results in asymptotically unbiased filtering estimates. However, the estimators based on the density approach require a great amount of computational burden compared with those based on the Taylor series approximations.[14]

Further, we constrain our analysis of Kalman filters on algorithms that are derived from the Taylor series expansions, i.e. we only deal with extended Kalman filters. For the exposition of the extended Kalman filter algorithms we work with the following general state space model.

[12] Note that these are then formally indexed with $t-1$, since we denote the actual information we are working with by the time index t.

[13] See, for example, Anderson and Moore (1979) and Tanizaki (1996).

[14] For a detailed comparison of different non-linear filtering algorithms see Tanizaki (1996).

Definition 2.6 (Non-Linear State Space Model) *We treat the non-linear filtering problem based on a state-space model with the measurement and transition equations being specified as*

$$\underset{g\times 1}{\mathbf{y}_t} = \mathbf{g}_t\left(\boldsymbol{\xi}_t, \boldsymbol{\varepsilon}_t(\boldsymbol{\psi}), \boldsymbol{\psi}\right), \text{ and} \tag{2.10}$$

$$\underset{k\times 1}{\boldsymbol{\xi}_t} = \mathbf{h}_t\left(\boldsymbol{\xi}_{t-1}, \boldsymbol{\eta}_t(\boldsymbol{\psi}), \boldsymbol{\psi}\right), \tag{2.11}$$

with $\mathbf{g}\left(\boldsymbol{\xi}_t, \boldsymbol{\varepsilon}_t(\boldsymbol{\psi}), \boldsymbol{\psi}\right)$ *and* $\mathbf{h}\left(\boldsymbol{\xi}_{t-1}, \boldsymbol{\eta}_t(\boldsymbol{\psi}), \boldsymbol{\psi}\right)$ *denoting the possible non-linear functional relationships. The error terms* $\boldsymbol{\varepsilon}_t(\boldsymbol{\psi})$ *and* $\boldsymbol{\eta}_t(\boldsymbol{\psi})$ *are assumed to follow the properties*

$$\mathbb{E}\begin{bmatrix}\boldsymbol{\varepsilon}_t(\boldsymbol{\psi})\\ \boldsymbol{\eta}_t(\boldsymbol{\psi})\end{bmatrix} = \mathbf{0}, \text{ and}$$

$$\mathbb{V}ar\begin{bmatrix}\boldsymbol{\varepsilon}_t(\boldsymbol{\psi})\\ \boldsymbol{\eta}_t(\boldsymbol{\psi})\end{bmatrix} = \begin{bmatrix}\mathbf{H}_t(\boldsymbol{\psi}) & \mathbf{0}\\ \mathbf{0} & \mathbf{Q}_t(\boldsymbol{\psi})\end{bmatrix}$$

as normally distributed random vectors.

In the case of the extended Kalman filters we approximate the non-linear equations $\mathbf{g}_t\left(\boldsymbol{\xi}_t, \boldsymbol{\varepsilon}_t, \boldsymbol{\psi}\right)$ and $\mathbf{h}_t\left(\boldsymbol{\xi}_{t-1}, \boldsymbol{\eta}_t, \boldsymbol{\psi}\right)$ around the conditional means and error terms $(\boldsymbol{\xi}_t, \boldsymbol{\varepsilon}_t) = \left(\boldsymbol{\xi}_{t|t-1}, \mathbf{0}\right)$ and $(\boldsymbol{\xi}_{t-1}, \boldsymbol{\eta}_t) = \left(\boldsymbol{\xi}_{t-1|t-1}, \mathbf{0}\right)$. The first-order Taylor series expansions around the vectors $(\boldsymbol{\xi}_t, \boldsymbol{\varepsilon}_t) = \left(\boldsymbol{\xi}_{t|t-1}, \mathbf{0}\right)$ result in the approximate expression

$$\mathbf{y}_t \approx \mathbf{g}_t\left(\boldsymbol{\xi}_{t|t-1}, \mathbf{0}, \boldsymbol{\psi}\right) + \mathbf{B}_{t|t-1}\left(\boldsymbol{\xi}_t - \boldsymbol{\xi}_{t|t-1}\right) + \mathbf{R}_{t|t-1}\boldsymbol{\varepsilon}_t, \tag{2.12}$$

$$\text{with } \mathbf{B}_{t|t-1} = \left.\frac{\partial \mathbf{g}_t(\boldsymbol{\xi}_t, \boldsymbol{\varepsilon}_t, \boldsymbol{\psi})}{\partial \boldsymbol{\xi}_t'}\right|_{(\boldsymbol{\xi}_t, \boldsymbol{\varepsilon}_t, \boldsymbol{\psi}) = (\boldsymbol{\xi}_{t|t-1}, \mathbf{0}, \boldsymbol{\psi})}, \text{ and}$$

$$\mathbf{R}_{t|t-1} = \left.\frac{\partial \mathbf{g}_t(\boldsymbol{\xi}_t, \boldsymbol{\varepsilon}_t, \boldsymbol{\psi})}{\partial \boldsymbol{\varepsilon}_t'}\right|_{(\boldsymbol{\xi}_t, \boldsymbol{\varepsilon}_t, \boldsymbol{\psi}) = (\boldsymbol{\xi}_{t|t-1}, \mathbf{0}, \boldsymbol{\psi})}$$

for the *measurement equation.* The *transition equation* is approximated around $(\boldsymbol{\xi}_{t-1}, \boldsymbol{\eta}_t) = \left(\boldsymbol{\xi}_{t-1|t-1}, \mathbf{0}\right)$ which results in:

$$\boldsymbol{\xi}_t \approx \mathbf{h}_t\left(\boldsymbol{\xi}_{t-1|t-1}, \mathbf{0}, \boldsymbol{\psi}\right) + \boldsymbol{\Phi}_{t|t-1}\left(\boldsymbol{\xi}_{t-1} - \boldsymbol{\xi}_{t-1|t-1}\right) + \mathbf{S}_{t|t-1}\boldsymbol{\eta}_t, \tag{2.13}$$

$$\text{with } \boldsymbol{\Phi}_{t|t-1} = \left.\frac{\partial \mathbf{h}_t(\boldsymbol{\xi}_{t-1}, \boldsymbol{\eta}_t, \boldsymbol{\psi})}{\partial \boldsymbol{\xi}_{t-1}'}\right|_{(\boldsymbol{\xi}_{t-1}, \boldsymbol{\eta}_t, \boldsymbol{\psi}) = (\boldsymbol{\xi}_{t-1|t-1}, \mathbf{0}, \boldsymbol{\psi})}, \text{ and}$$

$$\mathbf{S}_{t|t-1} = \left.\frac{\partial \mathbf{h}_t(\boldsymbol{\xi}_{t-1}, \boldsymbol{\eta}_t, \boldsymbol{\psi})}{\partial \boldsymbol{\eta}_t'}\right|_{(\boldsymbol{\xi}_{t-1}, \boldsymbol{\eta}_t, \boldsymbol{\psi}) = (\boldsymbol{\xi}_{t-1|t-1}, \mathbf{0}, \boldsymbol{\psi})}.$$

The *state space model* of equations (2.12) and (2.13) can further be stated as

$$\begin{aligned}
\mathbf{y}_t &\approx \mathbf{a}_t + \mathbf{B}_{t|t-1}\boldsymbol{\xi}_t + \mathbf{R}_{t|t-1}\boldsymbol{\varepsilon}_t \\
\boldsymbol{\xi}_t &\approx \mathbf{c}_t + \boldsymbol{\Phi}_{t|t-1}\boldsymbol{\xi}_{t-1} + \mathbf{S}_{t|t-1}\boldsymbol{\eta}_t, \\
&\quad \text{with } \mathbf{a}_t = \mathbf{g}_t\left(\boldsymbol{\xi}_{t|t-1}, \mathbf{0}, \boldsymbol{\psi}\right) - \mathbf{B}_{t|t-1}\boldsymbol{\xi}_{t|t-1}, \text{ and} \\
&\quad \mathbf{c}_t = \mathbf{h}_t\left(\boldsymbol{\xi}_{t-1|t-1}, \mathbf{0}, \boldsymbol{\psi}\right) - \boldsymbol{\Phi}_{t|t-1}\boldsymbol{\xi}_{t-1|t-1}
\end{aligned}$$

following the treatment of the linear state space model of equations (2.1) and (2.2). For the approximated state space model we are able to derive the following modification of the linear Kalman filter algorithm:[15]

$$\begin{aligned}
\boldsymbol{\xi}_{t|t-1} &= \mathbf{h}_t\left(\boldsymbol{\xi}_{t-1|t-1}, \mathbf{0}, \boldsymbol{\psi}\right), \\
\Sigma_{t|t-1} &= \boldsymbol{\Phi}_{t|t-1}\Sigma_{t-1|t-1}\boldsymbol{\Phi}'_{t|t-1} + \mathbf{S}_{t|t-1}\mathbf{Q}_t\mathbf{S}'_{t|t-1}, \\
\mathbf{v}_t &= \mathbf{y}_t - \mathbf{g}_t\left(\boldsymbol{\xi}_{t|t-1}, \mathbf{0}, \boldsymbol{\psi}\right), \\
\mathbf{F}_{t|t-1} &= \mathbf{B}_{t|t-1}\Sigma_{t|t-1}\mathbf{B}'_{t|t-1} + \mathbf{R}_{t|t-1}\mathbf{H}_t\mathbf{R}'_{t|t-1}, \\
\mathbf{K}_t &= \left(\mathbf{B}_{t|t-1}\Sigma_{t|t-1}\right)'\mathbf{F}^{-1}_{t|t-1}, \\
\boldsymbol{\xi}_{t|t} &= \boldsymbol{\xi}_{t|t-1} + \mathbf{K}_t\mathbf{v}_t, \text{ and} \\
\Sigma_{t|t} &= \Sigma_{t|t-1} - \mathbf{K}_t\mathbf{F}_{t|t-1}\mathbf{K}'_t.
\end{aligned}$$

This system of equations can recursively be implemented in analogy to the algorithm in the linear case presented in the previous section.

2.3 Parameter Estimation

The overview of the Kalman filter estimation procedure in figure 2.1 shows the likelihood function and the maximization procedure as the last relevant steps. We start to examine the appropriate *log-likelihood function* within our framework.[16] Let $\boldsymbol{\psi} \in \Psi \subset \mathbb{R}^n$ be the vector of the n unknown parameters, which various system matrices of the state space model formulated in equations (2.1) and (2.2), or (2.12) and (2.13), as well as the variances of the measurement errors depend on. The likelihood function of the state space model is given by the joint density of the observational data $\mathbf{y} = (\mathbf{y}_T, \mathbf{y}_{T-1}, \cdots, \mathbf{y}_1)$

[15] See Tanizaki (1996, ch. 3)

[16] A brad treatment of maximum likelihood estimation with time-series applications can be found, for example, in Stuart and Ord (1991), Hamilton (1994b), and Gourieroux and Monfort (1995b).

$$l(\mathbf{y};\psi) = p(\mathbf{y}_T, \mathbf{y}_{T-1}, \ldots, \mathbf{y}_1; \psi)$$

which reflects how likely it would have been to have observed the data if ψ were the true values for the parameters. Using the definition of conditional probability, we can split the likelihood up into conditional densities using *Bayes's theorem* recursively and write the joint density as the product of conditional densities

$$l(\mathbf{y};\psi) = p(\mathbf{y}_T|\mathbf{y}_{T-1}, \ldots, \mathbf{y}_1; \psi) \cdot \ldots \cdot p(\mathbf{y}_t|\mathbf{y}_{t-1}, \ldots, \mathbf{y}_1; \psi) \cdot \ldots \cdot p(\mathbf{y}_1; \psi), \tag{2.14}$$

where we approximate the initial density function $p(\mathbf{y}_1;\psi)$ by $p(\mathbf{y}_1|\mathbf{y}_0;\psi)$. Moreover, since we deal with a model with a *Markovian* structure according to the state space model, i.e. future values of $\mathbf{y}_u$, with $u > t$, only depend on $(\mathbf{y}_t, \mathbf{y}_{t-1}, \ldots, \mathbf{y}_1)$ through the current value $\mathbf{y}_t$, the expression in equation (2.14) is equivalent to conditioning on only the last observed vector of observations, i.e. it reduces to

$$l(\mathbf{y};\psi) = p(\mathbf{y}_T|\mathbf{y}_{T-1};\psi) \cdot \ldots \cdot p(\mathbf{y}_t|\mathbf{y}_{t-1};\psi) \cdot \ldots \cdot p(\mathbf{y}_1;\psi). \tag{2.15}$$

For our purpose of estimating the parameter vector ψ given the data $\mathbf{y}$ and the structural form of the specified state space model, we use the approach of Schweppe (1965) known as the *prediction error decomposition of the likelihood function* to be maximized with respect to ψ. That is, we express the likelihood function in terms of the prediction errors; since the variance of the prediction error $\mathbf{v}_t = \mathbf{y}_t - \mathbb{E}[\mathbf{y}_t|\mathcal{F}_{t-1}]$ is the same as the conditional variance of $\mathbf{y}_t$, i.e.

$$\mathbb{C}ov[\mathbf{v}_t|\mathcal{F}_{t-1}] = \mathbb{C}ov[\mathbf{y}_t|\mathcal{F}_{t-1}],$$

we are able to state the density function. The general density function $p(\mathbf{y}_t|\mathbf{y}_{t-1};\psi)$ is given by the Gaussian distribution with conditional mean $\mathbb{E}[\mathbf{y}_t|\mathcal{F}_{t-1}] = \mathbf{a} + \mathbf{B}\boldsymbol{\xi}_{t|t-1}$ and the conditional variance-covariance matrix $\mathbb{C}ov[\mathbf{y}_t|\mathcal{F}_{t-1}] = \mathbf{F}_{t|t-1}$, which takes the form:

$$p(\mathbf{y}_t|\mathbf{y}_{t-1};\psi) = \frac{1}{\sqrt{2\pi\left|\mathbf{F}_{t|t-1}\right|}} e^{-\frac{1}{2}\mathbf{v}_t'\mathbf{F}_{t|t-1}\mathbf{v}_t} \tag{2.16}$$

using the substitution $\mathbf{v}_t = \mathbf{y}_t - \mathbb{E}[\mathbf{y}_t|\mathcal{F}_{t-1}]$. Taking the logarithm of equation (2.16) yields the expression

$$\ln p(\mathbf{y}_t|\mathbf{y}_{t-1};\psi) = -\frac{k}{2}\ln 2\pi - \frac{1}{2}\ln\left|\mathbf{F}_{t|t-1}\right| - \frac{1}{2}\mathbf{v}_t'\mathbf{F}_{t|t-1}\mathbf{v}_t,$$

which gives us the whole *log-likelihood function* in the innovation form as the sum

$$L(\mathbf{y};\boldsymbol{\psi}) = -\frac{1}{2}\sum_{t=1}^{T} k\ln(2\pi) + \ln\left|\mathbf{F}_{t|t-1}\right| + \mathbf{v}_t'\mathbf{F}_{t|t-1}\mathbf{v}_t \tag{2.17}$$

instead of the product in equation (2.14).

In order to estimate the unknown parameters from the log-likelihood function in equation (2.17) we need to choose an appropriate *optimization algorithm* to minimize $L(\mathbf{y};\boldsymbol{\psi})$ with respect to $\boldsymbol{\psi}$. Given the simplicity of the innovation form of the likelihood function, we can find the values for the conditional expectation $\mathbb{E}[\mathbf{y}_t|\mathcal{F}_{t-1}]$ and conditional variance-covariance matrix $\mathbb{C}ov[\mathbf{y}_t|\mathcal{F}_{t-1}]$ for every given parameter vector $\boldsymbol{\psi}$ using the Kalman filter algorithm. Thus, we are able to compute the value of the conditional log-likelihood function numerically. This can then be used in an optimization algorithm of the log-likelihood function $L(\mathbf{y};\boldsymbol{\psi})$ in order to calculate the maximum likelihood estimates $\widehat{\boldsymbol{\psi}}_{ML}$ of the parameter vector $\boldsymbol{\psi}$:

$$\widehat{\boldsymbol{\psi}}_{ML} = \underset{\boldsymbol{\psi}\in\Psi}{\arg\max} L(\mathbf{y};\boldsymbol{\psi}), \tag{2.18}$$

where $\Psi \subset \mathbb{R}^n$ denotes the parameter space. In the case of implementing extended Kalman filter algorithms where we use approximations to the conditional normal distribution of equation (2.16) the log-likelihood of equation (2.18) can be used to yield quasi maximum likelihood parameter estimates.[17] However, the general statistical properties of such estimates must be studied separately and are assessed in detail by White (1982), Gallant and White (1988), and White (1994).

In the numerical implementation to obtain the optimal parameter estimates we actually minimize the log-likelihood function of equation (2.18) including its opposite sign. In infinite precision the necessary conditions for a local minimum $\widehat{\boldsymbol{\psi}}_{ML}$ of $L(\mathbf{y};\boldsymbol{\psi})$ are defined by the conditions for the gradient:

$$\nabla L\left(\mathbf{y};\widehat{\boldsymbol{\psi}}_{ML}\right) = 0, \tag{2.19}$$

assuming the relevant partial derivatives exist. A sufficient condition is that the matrix of second order derivatives exists and is positive definite at the optimal parameter estimates $\widehat{\boldsymbol{\psi}}_{ML}$, i.e.

[17] For an exploration of quasi (or pseudo) maximum likelihood estimation methods see, for example, Gourieroux, Monfort, Renault, and Trognon (1984).

$$\nabla^2 L\left(\mathbf{y};\widehat{\boldsymbol{\psi}}_{ML}\right) \succ 0 \tag{2.20}$$

is needed.

In order to estimate the model parameters $\boldsymbol{\psi}_{ML}$ we decided to choose a quasi-Newton method, which is similar to the methods used by Lund (1997) and Nunes and Clewlow (1999) based on Dennis and Schnabel (1996). The iteration rule for *Newton's method for unconstrained minimization* is given by the parameter estimates at the ith iteration step of

$$\boldsymbol{\psi}_{i+1} = \boldsymbol{\psi}_i + \mathbf{s}_i$$

where the variable $\mathbf{s}_i$ is found by solving

$$\nabla^2 L\left(\mathbf{y};\boldsymbol{\psi}_i\right)\cdot\mathbf{s}_i = -\nabla L\left(\mathbf{y};\boldsymbol{\psi}_i\right).$$

For the construction of $\nabla L\left(\mathbf{y};\boldsymbol{\psi}_i\right)$ and $\nabla^2 L\left(\mathbf{y};\boldsymbol{\psi}_i\right)$ we use numerical approximations. We found that the available closed-form solutions[18] do not lead to superior results in that they accumulate numerical errors in their required extensive calculations while asking for multiples of computer time. In cases where $\nabla^2 L\left(\mathbf{y};\boldsymbol{\psi}_i\right)$ is not positive definite Dennis and Schnabel (1996, sec. 5.5) suggest a modification of the modified Cholesky factorization. This algorithm changes the model when necessary so that it has a unique minimizer and uses this minimizer to define the Newton step. Specifically, at each iteration we take steps of the form

$$\mathbf{s}_i = -\left[\nabla^2 L\left(\mathbf{y};\boldsymbol{\psi}_i\right) + \mu_i\mathbf{I}\right]^{-1}\cdot\nabla L\left(\mathbf{y};\boldsymbol{\psi}_i\right)$$

by adding small parts μ_i of an identity matrix $\mathbf{I}$. Therein, μ_i is ideally not much larger than the smallest μ that will make $\nabla^2 L\left(\mathbf{y};\boldsymbol{\psi}_i\right) + \mu\mathbf{I}$ positive definite and reasonably well conditioned.

By now the optimization algorithm implies a full quasi-Newton step which will be taken whenever possible. In addition, we implement a backtracking line search framework for the step length λ_i. For properly chosen steps Dennis and Schnabel (1996, sec. 6.3) show that the algorithm ensures global convergence. Thereby, the optimization algorithm we use is a globally convergent modification of Newton's method. Its iteration rule of the optimization process can be represented as

$$\boldsymbol{\psi}_{i+1} = \boldsymbol{\psi}_i - \lambda_i\cdot\left[\nabla^2 L\left(\mathbf{y};\boldsymbol{\psi}_i\right) + \mu_i\mathbf{I}\right]^{-1}\cdot\nabla L\left(\mathbf{y};\boldsymbol{\psi}_i\right),$$

where $\boldsymbol{\psi}_i$ are the parameter estimates at the ith iteration step.

[18] As given, for example, in Harvey (1989, ch. 3.4).

Moreover, looking at the bottom part of the flowchart in figure 2.1, we need to discuss how to terminate our optimization algorithm in finite precision, i.e. we further have to examine the conditions stated in equations (2.19) and (2.20).[19] Although $\nabla L(\mathbf{y};\boldsymbol{\psi}) = 0$ can also occur at a maximum or saddle point, our globalizing strategy and our method of perturbing the model Hessian to be positive definite make convergence impossible to maxima and saddle points. Therefore, we consider $\nabla L(\mathbf{y};\boldsymbol{\psi}) = 0$ to be a necessary and sufficient condition for $\boldsymbol{\psi}$ to be a minimizer of $L(\mathbf{y};\boldsymbol{\psi})$.

To test whether $\nabla L(\mathbf{y};\boldsymbol{\psi}) = 0$, a test such as

$$|\nabla L(\mathbf{y};\boldsymbol{\psi})| \leq \varepsilon, \tag{2.21}$$

with a tolerance level of, for example, $\varepsilon = 10^{-4}$, is inadequate, because it is strongly dependent on the scaling of both $L(\mathbf{y};\boldsymbol{\psi})$ and $\boldsymbol{\psi}$. A direct modification of equation (2.21) is to define the relative gradient of $L(\mathbf{y};\boldsymbol{\psi})$ at $\boldsymbol{\psi}$ by

$$\textbf{rel grad } (\boldsymbol{\psi})_i = \lim_{\delta \to 0} \frac{\frac{L(\mathbf{y};\boldsymbol{\psi}+\delta\iota_i) - L(\mathbf{y};\boldsymbol{\psi})}{L(\mathbf{y};\boldsymbol{\psi})}}{\frac{\delta}{\boldsymbol{\psi}_i}} = \frac{\nabla L(\mathbf{y};\boldsymbol{\psi})_i \boldsymbol{\psi}_i}{L(\mathbf{y};\boldsymbol{\psi})}.$$

With the given relative gradient we perform the test

$$\max_{1 \leq i \leq n} |\textbf{rel grad } (\boldsymbol{\psi})_i| \leq \varepsilon \tag{2.22}$$

which is independent of any change in the units of $L(\mathbf{y};\boldsymbol{\psi})$ and $\boldsymbol{\psi}$. However the idea of relative change in $\boldsymbol{\psi}_i$ and $L(\mathbf{y};\boldsymbol{\psi})$ breaks down if $\boldsymbol{\psi}_i$ or $L(\mathbf{y};\boldsymbol{\psi})$ happen to be close to zero. This problem can be fixed by altering the test of equation (2.22) to

$$\max_{1 \leq i \leq n} \left| \frac{\nabla L(\mathbf{y};\boldsymbol{\psi})_i \max(|\boldsymbol{\psi}_i|, typ(\boldsymbol{\psi}_i))}{\max(|L(\mathbf{y};\boldsymbol{\psi})|, typ(L(\mathbf{y};\boldsymbol{\psi})))} \right| \leq \varepsilon, \tag{2.23}$$

where $typ(\mathbf{x})$ denotes an estimate of a typical magnitude of $\mathbf{x}$. Thus, we use the test stated in equation (2.23) in the Kalman filter applications.

[19] See Dennis and Schnabel (1996, ch. 7.2).

Part II

Pricing Equities

3

Introduction and Survey

3.1 Opening Remarks

In part II we investigate the pricing of equities. Within this asset class we draw our attention to the case of closed-end funds. We first describe the specific characteristics of closed-end funds and their motivating features for financial research, especially for issues of pricing. Thereupon, we develop a valuation model in chapter 4 by means of contingent claim pricing techniques attempting to capture the certain financial characteristics of closed-end funds. The stochastic pricing model especially takes into account both the price risk of the funds as well as their risk associated with altering discounts. Chapter 5 describes a possible implementation of the pricing model and introduces the estimation techniques used in the further analysis. For the empirical adaptation of the valuation model we use a sample of closed-end equity funds that invest in emerging markets and are traded on the New York Stock Exchange. The statistical inferences are based on a historical sample for the five-year-period of 1993 to 1997 using maximum likelihood estimation based on a Kalman filter algorithm. We estimate the relevant parameters and validate our model. In chapter 6 we are able to infer insights into two potential applications for investors. First, we test the forecasting power of the pricing model to predict closed-end fund market prices. Second, based on information that is revealed in the valuation model, we implement portfolio strategies using trading rules. The results on these suggested applications indicate that our pricing model generates valuable investment information. Finally, we summarize and interpret our theoretical and empirical findings in chapter 7.

3.2 Closed-End Funds: Survey and Hypotheses

In the categories of asset classes closed-end funds lie in between the class of common equity and open-end funds, which are generally known as mutual funds. In thus, they exhibit features of both asset classes which are worthwhile examining: (i) Closed-end funds are companies whose operations resemble those of any business corporation. Their shares are regularly traded on organized exchanges as any other publicly traded corporation. Closed-end funds only differ because their corporate business largely consists of investing funds in the securities of other entities and managing these investment holdings for income and profit. The important characteristic which makes closed-end funds unique among other joint-stock companies is that they provide simultaneous price quotations for both their stocks and their underlying investment portfolio. Especially, the share price behavior often results in prices being different from the value of their underlying investments as we will see further on. (ii) Closed-end funds are investment companies that provide investment management and bookkeeping services to investors. Especially, investors who do not have the time or expertise to manage their own funds rely on the services of investment companies. To qualify as such in the United States their activities are regulated by the investment Company Act of 1940 and by the 1950 amendments to the act.[1] The two major types of investment companies in the United States are open-end and closed-end investment companies. Open-end funds continuously issue and redeem ownership shares which are not traded on a secondary market or organized exchange. Instead investors purchase shares from the company and redeem shares by selling them back to the company. In thus, the equity capital and assets of a mutual fund are increased when shares are sold and reduced when shares are repurchased. The second class of investment companies are closed-end funds whose capitalization or outstanding number of shares is fixed or "closed". This implies that the supply of closed-end fund shares is inelastic, i.e. the price of the shares is a function of the supply and demand for the shares on the market. Therefore, there is only an indirect link with the value of the underlying investment portfolio, the so-called net asset value, corresponding to each share.

Combining these two aspects, the interesting feature of closed-end funds in comparison to common joint-stock companies and mutual funds is, that the market value determined on an organized exchange differs dynamically over time from the value of the underlying investment portfolio by a discount or a premium. This pricing phenomenon contradicts the well-known efficient mar-

[1] See, for example, Anderson and Born (1992).

ket hypothesis put forward by Fama (1970) and Fama (1991) which states that securities markets are expected to be informationally efficient. As noted by Malkiel (1977), the market for closed-end funds provides a "startling counter-example to the general rule".[2] He investigates several real world obstacles that could prevent the parity of the closed-end funds share prices and the net asset values. As a result he finds that the considered hypotheses can only account for a fraction in the observed discounts. Especially, the tax liability hypothesis of unrealized capital gains gives support to explain discount values of up to six percent. However, US domestic equity closed-end funds have traded at an average discount of around ten percent over the last thirty years.[3]

The various proposed hypotheses aiming at an explanation of the puzzle with the existing and observable closed-end fund discounts have grown to a wide body.[4] Broadly, the research put forward can be categorized into the standard economic theories and the behavioral explanations. The *standard economic theories* attempt to explain the observable premia within the efficient markets framework. At first, there are explanations that aim at the possibility that the closed-end funds' underlying portfolio values may be overestimated. This could be the case in that the net asset values do not reflect the capital gains tax that must be paid by the closed-end funds if appreciated assets in the fund are sold.[5] A further theory is given by the restricted stock hypothesis which states that substantial holdings of letter stock could be overvalued in the calculation of the net asset value.[6] Second, agency costs could create closed-end fund discounts if management expenses and fees are too high or if future portfolio management and performance is expected to be inadequate.[7] A last group of hypotheses focuses on various forms of market segmentation. Especially, market segmentation arises internationally when US investors diversify into inaccessible, less well regulated and under-researched

---

[2] Malkiel (1977, p. 847).

[3] See Swaminathan (1996).

[4] For a rich review of the evolving literature on closed-end funds see, for example, Anderson and Born (1992), Dimson and Minio-Kozerski (1999) and Anderson and Born (2002) .

[5] See, for example, Malkiel (1977), Lee, Shleifer, and Thaler (1990), Pontiff (1995), and Bierman and Swaminathan (2000).

[6] See the studies by Malkiel (1977) and Lee, Shleifer, and Thaler (1990).

[7] See, for example, Roenfeldt and Tuttle (1973), Ingersoll (1976), Malkiel (1977), and Thompson (1978).

foreign markets.[8] However, these standard arguments, even when considered together, do not explain all of the closed-end fund puzzle.[9]

The observed price differences with closed-end funds are even more interesting since closed-end funds are assets of high transparency in the sense that the companies underlying fundamental values are quite easy to determine compared to common equity. Instead, traded on organized exchanges the market prices of the closed-end funds seem to reflect more the supply and demand for these shares than the value of their underlying assets. The latter would be expected in a rational and efficient market. This brings us to the *second* line of closed-end fund literature which concentrates on behavioral explanations of the observable discounts. Here, the most prominent theory is the investor sentiment hypothesis proposed by Lee, Shleifer, and Thaler (1991).[10] The proposition is that fluctuations in premia are driven by changes in individual investor sentiment. By finding that closed-end fund premia are a measure of the sentiment of individual investors they conclude that a changing sentiment makes the funds riskier than their underlying portfolios and such causes the average underpricing of closed-end funds relative to their fundamentals. Thus, if the dynamic behavior of closed-end fund discounts reflects investor sentiment and is not conformable to rational asset pricing models, then there may be trading possibilities for investors on exploiting closed-end fund discounts. However, Swaminathan (1996) shows that the negative correlation between changes in discounts and small firm returns reported by Lee, Shleifer, and Thaler (1991) can be explained by the ability of the discounts to predict future small firm returns which are also affected by investor sentiment. He further provides evidence that information contained in the discounts may be related to fundamental explanations which suggests the possibility of a rational explanation for the described negative correlation. This leads Chordia and Swaminathan (1996) to build a noisy rational expectations model which includes market imperfections[11] instead of the existence of rational and irrational noisy traders as in the irrational asset pricing model of Lee, Shleifer, and Thaler (1991). Their model is capable of creating closed-end fund discounts endogenously within a rational setting which brings us back to the standard

[8] Especially, see the studies by Bonser-Neal, Brauer, Neal, and Wheatley (1990) and Diwan, Errunza, and Senbet (1995).

[9] See Lee, Shleifer, and Thaler (1991).

[10] The hypothesis is based on the noise trader model of De Long, Shleifer, Summers, and Waldmann (1990) who argue that this concept can explain both the persistent and variable premia on closed-end funds as well as the creation of such funds.

[11] They assume market segmentation and imperfect information.

economic explanations. They identify further market imperfections that may prevent investors from arbitrage gains using information on closed-end fund discounts.[12] The rational agent economy of Spiegel (1999) demonstrates that if agents have finite lives and asset supplies vary randomly over time, there emerges a systematic pattern of closed-end fund discounts as an equilibrium. Besides a number of empirical predictions for closed-end fund regularities this model can produce initial premia on closed-end funds. These, however, are only attainable as a randomly realized state of nature where it is not clear why one should expect this to happen ex ante.[13] The dynamic model of Arora, Ju, and Ou-Yang (2002) predicts a closed-end fund to start selling at a premium that can turn into a discount later. This model is formulated similar to Kyle (1985), endogenizes the price formation, and revives the issue of agency problems in closed-end fund management in that it relies on agency costs and information sharing.

An empirical study that examines the relevance of both, the standard theories as well as the hypothesis of investor sentiment, is provided by Gemmill and Thomas (2002). For a sample of UK closed-end funds they find support for a rational basis for the existence of a persistent discount which is considered to be driven by management expenses and asymmetric arbitrage. Further, they confirm that short and medium-term fluctuations of the premia are related to investor sentiment while they reject the hypothesis that noise is the cause of the observable discounts.

We currently see various theories of the determination and the variation of closed-end fund discounts available with no distinct prevailing hypothesis. The consensus of the different aspects is basically revealed in the investors' supply and demand behavior and is finally disclosed in the realized market prices of the closed-end fund shares. Upon the fact that dynamically changing premia and discounts predominantly exist in the market of closed-end funds we develop a valuation model to price closed-end fund shares. The objective is to provide a theoretical and empirical analysis of the dynamic price behavior of closed-end funds by incorporating the major types of risk associated with an investment in closed-end funds. These types of risk are considered to be the common price risk due to changing underlying portfolio values as well as the risk associated with altering values of the premia. In the following chapter

---

[12] The main issues are found to be related to securities regulations, fiduciary responsibilities, and free-rider problems.

[13] Khorana, Wahal, and Zenner (2002) find that closed-end funds on average announce rights offerings when the funds trade at a premium. Over the course of the offering this premium turns into a discount.

we derive a two-factor pricing model that utilizes the net asset values of the closed-end funds and a dynamically changing premium as state variables.

4

Valuation Model

4.1 Characteristics of Closed-End Funds

The capital markets for closed-end funds provide two important prices for financial analysis: The market value of the closed-end funds' shares as determined on organized exchanges, and the net asset value of their foreign assets which is reported by the investment companies.

Definition 4.1 *The net asset value (NAV) or the 'book value of shareholder's equity' for time t is given by*

$$NAV_t = (\textit{Total Assets} - \textit{Total Liabilities}) \,/\, (\textit{Shares Outstanding})$$

which generally differs from the market price P_t of the closed-end funds.[1] *This price difference per share is generally quantified by*

$$PREM_t = \ln (P_t / NAV_t) \tag{4.1}$$

which defines the empirical premium.

The empirical premium serves as a characteristic number for closed-end funds and is regularly published in the financial press and contained in financial data bases. From its definition the empirical premia can realize positive and negative values.[2] This is confirmed empirically whereas we rarely find values for the premia which are zero or even close to zero. This partly stems from the fact compared to open-end mutual funds, that closed-end fund investors agree to trade their shares at prices differing from the market value of the closed-end funds assets on organized exchanges. In order to liquidate a

[1] The net asset value and the market price are generally stated per share.

[2] Negative empirical premia correspond to the case of closed-end funds trading at *discounts*.

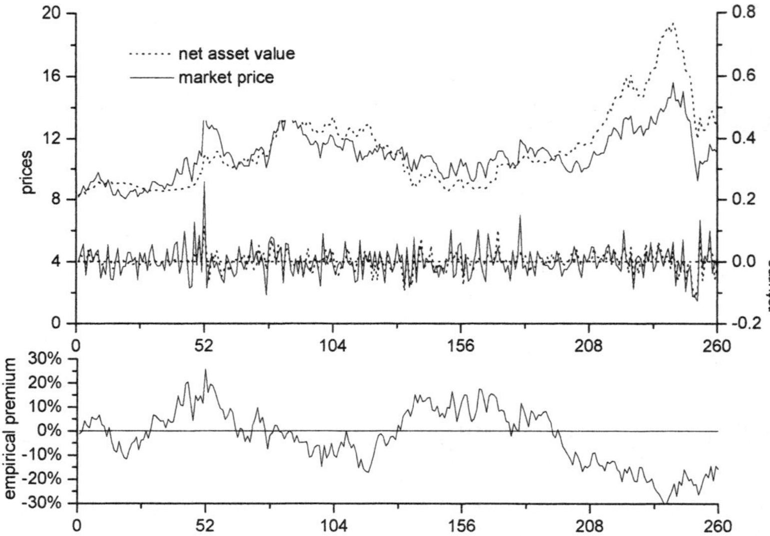

Fig. 4.1. ROC Taiwan Fund (ROC)

holding in a fund, the investors need to sell their shares to other investors for the market price instead of selling and redeeming them at or near their net asset value as with open-end funds. Thus, closed-end funds are endowed with an inelastic supply of shares.

Considering the closed-end funds in our sample[3] we can identify three different patterns of empirical premia behavior over time: (i) Premia which change from positive to negative values over the examined period of time, (ii) mainly positive premia over time, and (iii) negative premia or discounts for the major part of the sample. Out of our examined sample we choose three representative funds for each case: The ROC Taiwan Fund (ROC), the Indonesia Fund (IF), and the GT Global Eastern Europe (GTF). The time-series properties of these funds are illustrated, respectively, in figures 4.1, 4.2 and 4.3.[4]

The figures show the closed-end fund prices for the market shares and the reported net asset values in the upper graph combined with a graph of the total returns to these prices. The lower graph displays the time-series of the empirical premia of the funds. In the case of the ROC Taiwan Fund (figure

[3] For the detailed description of the data set see section 5.1 on page 59.

[4] Further information on the empirical premia of our sample is provided in table 5.2 on page 62.

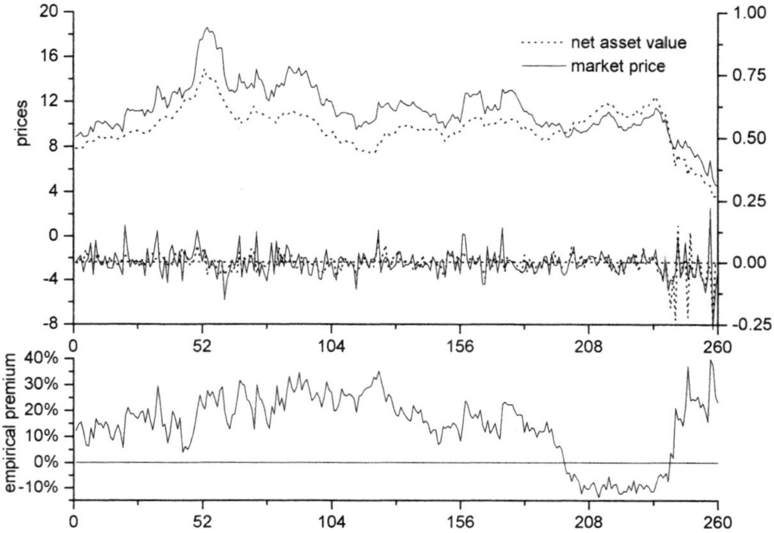

Fig. 4.2. Indonesia Fund (IF)

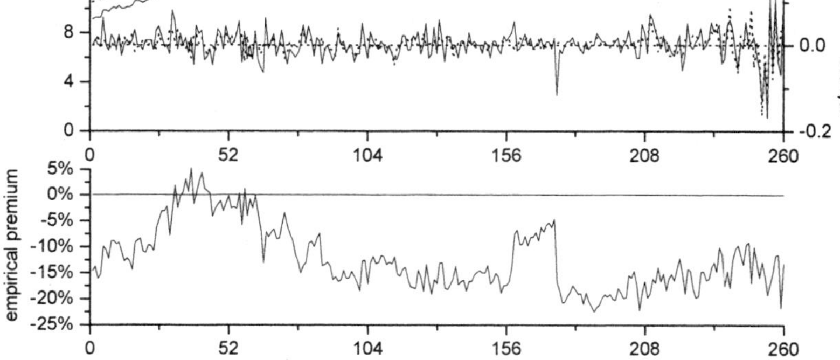

Fig. 4.3. GT Global Eastern Europe (GTF)

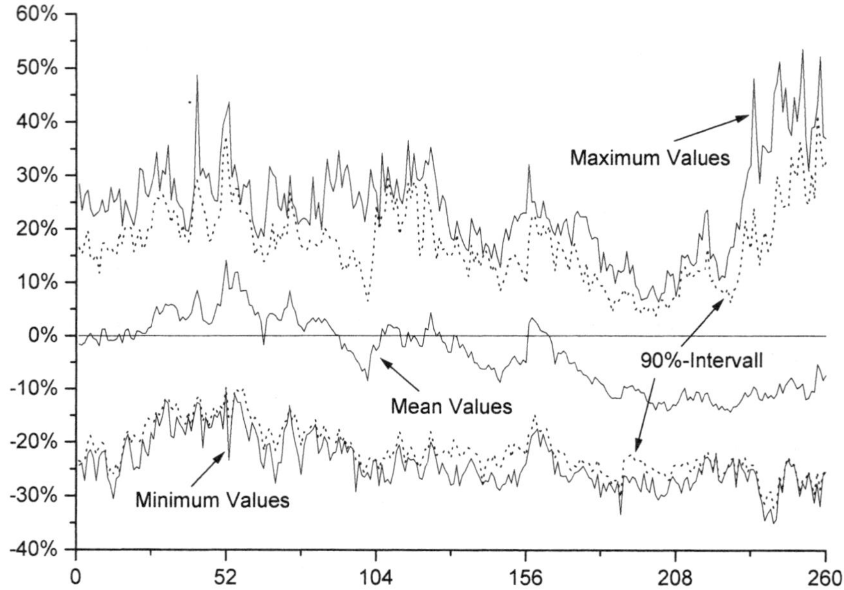

Fig. 4.4. Time Series of the Empirical Premia

4.1) the premium oscillates between positive and negative values. Looking at figures 4.2 and 4.3, we see that the Indonesia Fund and the GT Global Eastern Europe, respectively, exhibit primarily positive and negative values for the premium. For all three funds, we further notice that the price returns show a higher volatility for the market prices than for the net asset values.

The visual examination of the three selected closed-end funds from the sample seems to be in line with the existing literature on closed-end fund valuation. The literature sees the empirical premia of the closed-end funds to be subject to wide fluctuations over time and by no means a constant fraction of the net asset value (see Lee, Shleifer, and Thaler (1991)). As we take a look at the values of the premia across all closed-end funds in our sample (see figure 4.4) over the examined period of time, we can confirm this view based on the whole sample. For the average premia we observe mainly positive values for the first half of the sample period with a peak of 13 percent at the end of the first year, which tends to continually fall to discount values of about 14 percent in the fourth and fifth year. Looking at the maximum positive and negative values, we notice a higher volatility for the positive premia whereas the discounts oscillate in a narrow band of minus 10 to 30 percent.

Moreover, for Sharpe and Sosin (1975) the fluctuations of the empirical premia appear to be mean-reverting. For the analyzed sample, they find a slight tendency for the premia to revert to a long-term mean discount value of approximately seven percent. In an examination of the relation between closed-end fund premia and returns, Pontiff (1995) finds no economically motivated explanation for discounted closed-end funds having higher expected returns than non-discounted closed-end funds, but rather accounts a mean-reverting nature of the empirical premia for this effect. For the market of closed-end funds in the UK, the study of Minio-Paluello (1998) shows the empirical tendency of premia to revert to their mean. Gasbarro, Johnson, and Zumwalt (2003) apply cointegration techniques to identify mean-reverting closed-end funds on a sample of equity and bond funds and provide evidence on the speed of reversions. They statistical significantly find that most closed-end funds exhibit mean-reverting tendencies caused by changes in both the share price and the net asset value. These results from the literature give rise to capturing this pricing phenomenon in a stochastic framework of closed-end fund share valuation which we lay out in the next section.[5]

4.2 Economic Foundation

The proposed model structure links the share prices of the closed-end funds to the value of their underlying portfolio invested in the originating markets by incorporating a dynamically changing premium under a stochastic framework. Before performing this linkage, we first specify the frictionless economy underlying our continuous-time valuation model.

Assumption 4.1 (Frictionless Economy) *We assume a risk-free asset at a constant interest rate r that is available to all market participants in the economy. The risky assets in the economy are traded on either of the following markets:*

(i) the primary market, i.e. the local home market of the countries where the closed-end fund portfolios are originally invested, and

(ii) the secondary market for the shares of the closed-end funds, i.e. the US host market.

[5] Further evidence on widely varying closed-end fund discounts that appear to be mean-reverting are found in Thompson (1978), Richards, Fraser, and Groth (1980), Anderson (1986), and Brauer (1988).

In the case of international closed-end funds traded in the US the primary market could, for example, be Brazil. Thereby, the closed-end fund shares and their underlying portfolio of foreign securities are priced in different market segments. In modeling the closed-end fund prices on the secondary markets we choose the net asset value to be the first model factor. This choice is motivated by the fact that the value of the component assets in the originating countries are fundamental to value closed-end funds on the advanced secondary markets. From the literature we know that the net asset value explains most of the return on closed-end fund share prices but not all of it.[6]

Assumption 4.2 (Net Asset Value) *For the primary market we model the price movements for the investment portfolio of the closed-end funds measured by the reported net asset value as geometric Brownian motion*[7]

$$dNAV_t/NAV_t = \mu dt + \sigma_X dW_{X,t},$$

as in Ingersoll (1976).

Corollary 4.3 *For the assumed price behavior of the net asset value we conveniently work with the stochastic dynamics of the log-transformed variable* $X_t = \ln NAV_t$ *given by*

$$dX_t = \left(\mu - \frac{1}{2}\sigma_X^2\right) dt + \sigma_X dW_{X,t}, \qquad (4.2)$$

where dX_t *is the continuous return on the net asset value portfolio.*

In addition to the price risk modeled by the first factor we introduce a *time continuous premium* π_t on the expected return of closed-end funds. How this dynamically modeled premium is related to the empirical premium will be clarified further on.[8] Such a premium is intended to capture the stochastic price behavior of the closed-end fund market prices that is superimposed on the dynamics of the underlying asset values. The stochastic specification of this second factor is motivated by the observed mean-reverting dynamic fluctuations of the empirical premia.

[6] For example, Chen, Kan, and Miller (1993) find that 72.8 percent is explained by the net asset value variance on a closed-end fund sample covering the period from 1965 to 1985.

[7] A stochastic process corresponding to an exponential growth of the state variable which is a widespread asset pricing assumption for modeling equities in financial economics.

[8] Specifically, see the description on page 65 for this relationship.

Assumption 4.4 (Dynamic Premium) *The dynamic premium is assumed to follow a stochastic mean-reverting specification of the form*

$$d\pi_t = \kappa (\theta - \pi_t) \, dt + \sigma_\pi dW_{\pi,t}, \tag{4.3}$$

which is a type of Ornstein-Uhlenbeck process. In this special case the continuous premium π_t is modeled with a volatility of σ_π and reverts with a speed of κ to its long-run mean θ. Further, the derived valuation model will allow for a correlation of the two model factors X_t and π_t as the standardized Brownian motions are set up as $dW_{X,t}dW_{\pi,t} = \rho dt$.

Now, we link the price of the closed-end fund shares that are traded on the secondary market both to the values of their underlying net asset value and the dynamic premium. We model the market value of the closed-end fund share prices, denoted by $P(X, \pi, t; \psi_M)$ with parameter vector ψ_M, as a derivative security dependent on the values of the NAV_t and the premium π_t. In analogy to the standard technique known from contingent claim pricing[9], we derive the following stochastic dynamics for $P(X, \pi, t; \psi_M)$ by applying Itô's formula

$$\frac{dP}{P} = \mu_P dt + \eta_X dW_X + \eta_\pi dW_\pi, \tag{4.4}$$

$$\begin{aligned} \text{with } \mu_P = \frac{1}{P} \Bigg[& \left(\mu - \frac{1}{2}\sigma_X^2 \right) \frac{\partial P(X, \pi, t; \psi_M)}{\partial X} + \kappa (\theta - \pi) \frac{\partial P(X, \pi, t; \psi_M)}{\partial \pi} \\ & + \frac{\partial P(X, \pi, t; \psi_M)}{\partial t} + \frac{1}{2}\sigma_X^2 \frac{\partial^2 P(X, \pi, t; \psi_M)}{\partial X^2} \\ & + \rho \sigma_X \sigma_\pi \frac{\partial^2 P(X, \pi, t; \psi_M)}{\partial X \partial \pi} + \frac{1}{2}\sigma_\pi^2 \frac{\partial^2 P(X, \pi, t; \psi_M)}{\partial \pi^2} \Bigg], \\ \eta_X = \frac{1}{P}\sigma_X & \frac{\partial P(X, \pi, t; \psi_M)}{\partial X}, \text{ and } \eta_\pi = \frac{1}{P}\sigma_\pi \frac{\partial P(X, \pi, t; \psi_M)}{\partial \pi}, \end{aligned}$$

where the time indices are temporarily dropped for legibility.

Taking a closer view on the model economy with its three specified markets, we further need to ensure the consistent pricing of the different assets across all markets. Therefore, we take the view of an US investor that compares returns of alternative investments denominated in home currency. In an equilibrium state of the modeled economy, investors are indifferent in holding either of the available assets according to their risk preferences. Using the *standard finance condition* for a market equilibrium with no-arbitrage opportunities, we specify the necessary expected yield above the risk-free rate as

[9] See, for example, Merton (1992).

$\mu - r = \sigma_X \lambda_X$ including the market price of risk λ_X.[10] This ensures that there are no arbitrage possibilities on the primary market to gain riskless profits. For the secondary market, where the closed-end fund shares are traded, investors require an expected return of $\mu_P = r + \eta_X \lambda_X + \pi + \eta_\pi \lambda_\pi$ including the dynamic premium π and its market price of risk λ_π.[11] Interpreting the latter equilibrium condition from a discounting perspective, we expect the implicit discount rate μ_P of the true terminal value to be high when the premium is positive. Contrarily, closed-end funds with negative values for the premium reveal investors' attitudes towards a low expected return. In addition, by incorporating both risk premia for the modeled risks η_X and η_π our valuation formula will contain both the price risk and the premium risk inherent, respectively, in changes of the NAV and varying values for the continuous premium.

4.3 Pricing Closed-End Fund Shares

Based on the outlined economic foundations of the valuation model we now derive the fair market price of closed-end fund shares. In a first step, we use the underlying equilibrium conditions in conjunction with the stochastic dynamics of the closed-end funds' share prices from equation (4.4). This yields the *fundamental partial differential equation* underlying our valuation model

$$\begin{aligned}
&\tfrac{1}{2}\sigma_X^2 \frac{\partial^2 P(X,\pi,t;\psi_M)}{\partial X^2} + \rho\sigma_X\sigma_\pi \frac{\partial^2 P(X,\pi,t;\psi_M)}{\partial X \partial \pi} + \tfrac{1}{2}\sigma_\pi^2 \frac{\partial^2 P(X,\pi,t;\psi_M)}{\partial \pi^2} \\
&+ \left(r - \tfrac{1}{2}\sigma_X^2\right) \frac{\partial P(X,\pi,t;\psi_M)}{\partial X} + \left(\kappa\left(\theta - \pi\right) - \sigma_\pi \lambda_\pi\right) \frac{\partial P(X,\pi,t;\psi_M)}{\partial \pi} + \frac{\partial P(X,\pi,t;\psi_M)}{\partial t} \\
&= \left(r + \pi\right) P,
\end{aligned} \tag{4.5}$$

which describes how the closed-end funds' share price P varies for different values of the state variables X and π as well as for changing trading time t. Additionally, to fully reflect the dynamic behavior of the closed-end funds' share price $P(X, \pi, t; \psi_M)$ we further need to specify the appropriate boundary condition for this security. To defer the relevant condition, we exploit the economic behavior of the closed-end funds at the fourth stage of their life cycle at time T.[12] According to Brauer (1984) and Brickley and Schallheim

---

[10] See, for example, Ingersoll (1987).

[11] The two market prices of risk, λ_X and λ_π, are treated as constants in our model.

[12] According to the nomenclature of Lee, Shleifer, and Thaler (1991).

(1985) the market values of the closed-end funds equal their net asset values with empirical regularity at the time of the funds liquidation or open-ending. From this fact we obtain the boundary condition $P(X,\pi,T;\psi_M) = e^{X_T}$ for our valuation model.

The next step in deriving the price of closed-end fund shares in our model is to solve the partial differential equation of equation (4.5). For the solution to this equation we consider the exponential affine form[13]

$$P(X,\pi,t;\psi_M) = e^{A(t)X+B(t)\pi+C(t)}, \tag{4.6}$$

with model parameters ψ_M and factor loadings $A(t)$, $B(t)$, and $C(t)$. Using the partial derivatives of the proposed solution

$$P_X = \frac{\partial P(X,\pi,t;\psi_M)}{\partial X} = AP, \qquad P_{XX} = \frac{\partial^2 P(X,\pi,t;\psi_M)}{\partial X^2} = A^2 P,$$

$$P_\pi = \frac{\partial P(X,\pi,t;\psi_M)}{\partial \pi} = BP, \qquad P_{\pi\pi} = \frac{\partial^2 P(X,\pi,t;\psi_M)}{\partial \pi^2} = B^2 P, \text{ and}$$

$$P_{X\pi} = \frac{\partial^2 P(X,\pi,t;\psi_M)}{\partial X \partial \lambda} = ABP,\ P_t = \frac{\partial P(X,\pi,t;\psi_M)}{\partial t} = (A_t X_t + B_t \pi_t + C_t)\,P$$

for the differential equation (4.5) results in

$$\begin{aligned} &\left[\tfrac{\partial A}{\partial t}\right] X_t + \left[\tfrac{\partial B}{\partial t} - \kappa B - 1\right] \pi_t + \left[\tfrac{1}{2}\sigma_X^2 A^2 + \rho\sigma_X\sigma_\pi AB + \tfrac{1}{2}\sigma_\pi^2 B^2 \right. \\ &\left. + \left(r - \tfrac{1}{2}\sigma_X^2\right) A + (\kappa\theta - \sigma_\pi\lambda_\pi)\, B + C_t - r\right] = 0, \end{aligned} \tag{4.7}$$

where we already separated for the variables.

In order for equation (4.7) to hold for every values of X_t and π_t, we need the expressions in square brackets to be zero. This leads us to a system of ordinary differential equations in $A(t)$, $B(t)$, and $C(t)$. For successively solving this system we make use of the original economic boundary condition that the market values of the closed-end funds equal their net asset values at the fourth stage of their life cycle which translates into

$$A(T) = 1, \text{ and } B(T) = C(T) = 0.$$

Making use of the first two conditions we obtain the following solutions for the functions $A(t)$ and $B(t)$:

$$\begin{aligned} A(t) &= 1, \text{ and} \\ B(t) &= -\frac{1}{\kappa}\left(1 - e^{-\kappa(T-t)}\right). \end{aligned}$$

[13] This type of solution is known from interest rate modeling; see, for example, Duffie and Kan (1996).

The last function of interest is $C(t)$. Solving the integral

$$\int_t^T C_\zeta d\zeta = C(T) - C(t)$$

in conjunction with the boundary condition $C(T) = 0$ yields:

$$\begin{aligned} C(t) = &- \left[\theta - \frac{1}{\kappa}\sigma_\pi\lambda_\pi + \frac{1}{\kappa}\rho\sigma_X\sigma_\pi - \frac{\sigma_\pi^2}{2\kappa^2}\right](T-t) \\ &+ \frac{1}{\kappa}\left[\theta - \frac{1}{\kappa}\sigma_\pi\lambda_\pi + \frac{1}{\kappa}\rho\sigma_X\sigma_\pi - \frac{\sigma_\pi^2}{\kappa^2}\right]\left(1 - e^{-\kappa(T-t)}\right) \\ &+ \frac{\sigma_\pi^2}{4\kappa^3}\left(1 - e^{-2\kappa(T-t)}\right). \end{aligned}$$

To further derive the market price of the closed-end fund shares, we now use the calculated functions $A(t)$, $B(t)$, and $C(t)$ along with the suggested solution in equation (4.6).

Solution 4.1 (Closed-End Funds' Market Price) *With the log-transformation of the closed-end funds' market price* $Y(X, \pi, t; \psi_M) = \ln P(X, \pi, t; \psi_M)$ *we finally obtain the general valuation formula*

$$\begin{aligned} Y(X, \pi, t; \psi_M) = &X_t - \frac{1}{\kappa}\left(1 - e^{-\kappa(T-t)}\right)\pi_t \\ &- \left(\theta - \frac{1}{\kappa}\sigma_\pi\lambda_\pi + \frac{1}{\kappa}\rho\sigma_X\sigma_\pi - \frac{\sigma_\pi^2}{2\kappa^2}\right)(T-t) \\ &+ \frac{1}{\kappa}\left(\theta - \frac{1}{\kappa}\sigma_\pi\lambda_\pi + \frac{1}{\kappa}\rho\sigma_X\sigma_\pi - \frac{\sigma_\pi^2}{\kappa^2}\right)\left(1 - e^{-\kappa(T-t)}\right) \\ &+ \frac{\sigma_\pi^2}{4\kappa^3}\left(1 - e^{-2\kappa(T-t)}\right) \end{aligned} \tag{4.8}$$

for the log-market price of closed-end funds with a maturity of T.

The derived valuation formula states that the log-market price Y_t should basically be equal to the log-value X_t of the underlying assets in that the market prices carry the price risk with a factor loading of $A(t) = 1$. Second, the closed-end funds' market prices are affected by changing values of the dynamic premium π_t. The realizations of the premia influence the market prices by a negative loading $B(t)$, i.e. closed-end funds with positive values for the dynamic premium trade at a discount. Finally, the valuation formula contains a time varying term $C(t)$ which depends on the maturity of the closed-end funds and the model parameters ψ_M. In the following chapter we motivate a possible implementation of the valuation model upon market data of closed-end funds.

5

First Empirical Results

5.1 Sample Data

For the empirical implementation of our pricing model we choose to select a broad market sample of emerging market closed-end funds.[1] The sample examined in this study exists of closed-end equity funds for a complete five year period from January 1993 to December 1997. All NYSE traded closed-end funds are included, if their date of issue lies before January 1993 in order to avoid post-offering pricing effects.[2] The data is collected on a weekly basis to yield a multiple time-series of 33 closed-end funds with 260 observations each. The market prices are obtained from the last trade on Fridays, and the net asset values are calculated by the funds on Thursdays balances. However, the net asset values are reported on Fridays after the NYSE closes which makes trading on the basis of these prices impossible, but are treated as being observed contemporaneously with the market prices.[3] Finally, the share prices and net asset values are adjusted for distributions including income dividends and capital gains. In table 5.1 we report the figures for the total returns and their correlation along with the net proceeds and the date of the initial public offering (IPO) of the funds.

The returns on the different closed-end funds show a wide range according to their core countries of investment. Funds invested in Brazil, for example, realize about 15 percent on their market prices, whereas investment portfolios in Thailand show a 18 percent loss on average over the five year period. The returns on the net asset values also show a wide range, but are not as volatile.

---

[1] I greatly thank *Lipper Analytical Services*, and especially Donald Cassidy for kindly providing the closed-end fund data.

[2] See, for example, Peavy (1990) and Weiss (1989).

[3] Thus, the empirical premia will include some undetermined amount of measurement error.

Table 5.1. Characteristics of the Closed-End Fund Sample

Fund Name	Sym-bol	IPO Date	Pro-ceeds[a]	P-Returns[b)c)] MEAN	STD	NAV-Returns[c)] MEAN	STD	Corr[d)]
Argentina Fund	AF	10/91	63.3	4.13	33.09	11.35	22.15	0.699
Asia Pacific Fund	APB	04/87	79.0	-0.09	33.60	3.42	20.84	0.554
Brazil Fund	BZF	03/88	138.4	14.44	38.55	18.33	37.32	0.700
Brazilian Equity Fund	BZL	04/92	63.4	15.35	45.91	17.65	41.04	0.635
Chile Fund	CH	09/89	64.0	8.06	27.11	12.59	17.52	0.674
China Fund	CHN	07/92	110.7	1.09	32.33	3.87	21.47	0.508
Clemente Glob Growth	CLM	06/87	54.9	10.05	16.62	10.24	11.17	0.500
Emerging Mexico Fund	MEF	10/90	55.0	-3.30	39.69	1.98	32.96	0.677
Emerg MKTS Telecom	ETF	06/92	115.3	10.03	27.77	13.09	16.05	0.605
First Israel Fund	ISL	10/92	68.7	2.33	27.32	3.93	18.33	0.503
First Philippine Fund	FPF	11/89	98.9	-2.09	28.88	-5.27	22.40	0.578
Greater China Fund	GCH	07/92	93.1	5.14	33.47	7.45	24.53	0.595
GT Global East Europe	GTF	03/90	224.6	14.42	23.97	14.12	17.59	0.560
India Growth Fund	IGF	08/88	55.0	-5.82	30.92	-2.09	24.48	0.477
Indonesia Fund	IF	03/90	63.6	-13.36	37.74	-15.54	30.11	0.486
Jakarta Growth Fund	JGF	04/90	54.9	-13.37	35.10	-14.74	29.07	0.517
Jardine Fleming China	JFC	07/92	94.1	-3.78	32.44	-0.75	22.48	0.513
Korea Fund	KF	08/84	54.9	-12.34	40.26	-13.14	32.71	0.581
Korean Investm Fund	KIF	02/92	45.8	-17.49	37.12	-23.43	34.66	0.613
Latin Amer Equity FD	LAQ	10/91	83.1	6.18	32.97	10.28	21.32	0.579
Latin Amer Investment	LAM	07/90	54.6	7.19	31.94	9.84	20.88	0.612
Latin Amer Discovery	LDF	06/92	80.2	14.83	35.38	12.04	28.47	0.602
M S Emerg MKTS	MSF	10/91	148.1	3.37	29.07	7.33	16.59	0.495
Malaysia Fund	MF	05/87	80.2	-9.37	31.69	-14.28	28.87	0.534
Mexico Equity&Income	MXE	08/90	70.4	5.17	36.83	11.05	27.83	0.672
Mexico Fund	MXF	06/81	112.1	-0.80	39.04	3.50	36.25	0.759
Roc Taiwan Fund	ROC	05/89	54.5	6.06	34.98	8.97	22.64	0.605
Scudder New Asia Fund	SAF	06/87	77.5	-1.46	29.98	1.92	19.13	0.572
Singapore Fund	SGF	07/90	54.9	0.33	30.85	-0.33	18.51	0.471
Taiwan Fund	TWN	12/86	23.9	7.39	35.87	10.12	25.83	0.493
Tempelt Emerg MKTS	EMF	02/87	106.4	12.67	29.09	14.56	16.46	0.334
Taih Capital Fund	TC	05/90	67.9	-16.77	35.26	-23.50	28.32	0.586
Taih Fund	TTF	02/88	105.7	-18.71	35.28	-27.18	31.46	0.570
		MEAN	82.3	0.89	33.03	2.04	24.83	0.571
		STD	37.5	9.91	5.40	12.52	7.11	0.084
		MAX	224.6	15.35	45.91	18.33	41.04	0.759
		MIN	23.9	-18.71	16.62	-27.18	11.17	0.334

Notes:
a) Net proceeds are in million $.
b) P-returns denote returns on the market prices of the closed-end fund shares.
c) Returns are annualized log-returns, calculated on a weekly basis, and given in percent.
d) Denotes the correlation coefficient between the P- and NAV-returns.

The standard deviation of the net asset value returns averages on 25 percent and is about one-half lower than the 33 percent on the market returns.[4] This empirical result is in line with Pontiff (1997) who finds that closed-end fund prices are more volatile than the underlying net asset value. Looking at the correlation between the two returns, we see a strong positive comovement with a correlation coefficient of 0.57 on average.

In table 5.2 we show the descriptive statistics for the empirical premia of each closed-end fund. Across all closed-end funds in the sample, we find an average discount of 3.25 percent which is in line with previous research on emerging market closed-end funds.[5] The range of premia is given by the spanning from a 20 percent discount to positive premia of 15 percent on average. However, only eight out of the 33 closed-end funds trade at positive premia on average. Especially, the four closed-end funds including the symbols IGF, IF, KF and EMF show premia that seem to be persistently on a level above zero over time. For these funds we observe positive values for at least 75 percent of the sample.

To finally characterize the sample data, we further analyze the time-series properties of the empirical premia. According to the described literature in section 4.1 the empirical premia show some evidence to behave in a mean-reverting manner. Therefore, we model the premia $PREM_t$ ad hoc by an Ornstein-Uhlenbeck process with a specification of the form

$$dPREM_t = \varphi\left(\chi - PREM_t\right)dt + \omega dW_{PREM,t}, \tag{5.1}$$

where the speed of mean-reversion is denoted by φ, the long-run mean given by χ, and ω stands for the volatility term. With this specification the empirical premia follow a Gaussian distribution which can be estimated using an exact maximum likelihood method for the discretized process.

The results of the empirical premia in our sample for the discretized stochastic specification of equation (5.1) are shown in table 5.3. The parameter estimates result in average values for the coefficient of mean-reversion of 3.9, for the long-run mean value of the empirical premia of -0.04 and an annual standard deviation of 28 percent. The mean-reversion coefficient shows a high fluctuation of values from 1.4 up to 9.8. The estimations for the long-run mean exhibit positive as well as negative values related to whether the closed-end fund trades mainly with a premium or a discount over the sample period.

[4] This excess price volatility compared to the fundamental volatility is consistent with the implications of the De Long, Shleifer, Summers, and Waldmann (1990) noise trader model.

[5] See, for example, Dimson and Minio-Kozerski (1999).

Table 5.2. Descriptive Statistics on the Closed-End Fund Premia

Fund	Empirical Premia[a]								
Name	MEAN	STD	MAX	UQ	MED	LQ	MIN	SK[b]	KU[b]
AF	-0.36	9.11	20.23	6.95	1.38	-7.60	-21.38	-0.26	2.18
APB	-0.28	13.57	28.51	11.26	2.14	-13.63	-23.30	0.02	1.71
BZF	-5.73	10.55	16.24	2.37	-4.05	-16.46	-28.65	-0.21	1.95
BZL	-5.77	12.00	26.68	3.69	-5.56	-16.99	-33.41	0.14	2.30
CH	-8.63	7.79	12.35	-4.58	-9.59	-14.20	-24.29	0.49	2.55
CHN	-1.30	13.57	30.51	9.45	0.97	-11.84	-29.51	-0.25	2.02
CLM	-20.46	5.96	-6.01	-15.60	-21.41	-25.64	-30.66	0.32	1.98
MEF	-5.20	14.81	34.15	2.46	-4.40	-19.11	-28.91	0.59	2.72
ETF	-7.13	12.07	23.31	2.63	-8.36	-18.43	-24.50	0.50	2.10
ISL	-6.79	12.67	23.39	1.41	-10.97	-16.23	-28.65	0.74	2.44
FPF	-18.14	6.38	8.12	-15.76	-19.26	-22.14	-30.56	1.46	5.94
GCH	-7.53	11.38	17.35	2.13	-7.13	-18.35	-30.68	0.01	1.86
GTF	-12.22	6.06	5.37	-8.63	-13.50	-16.65	-22.48	0.74	2.83
IGF	8.10	10.49	43.75	14.49	7.21	0.31	-18.23	0.37	3.44
IF	14.57	12.41	39.97	23.31	16.82	9.95	-13.80	-0.74	2.79
JGF	2.77	9.44	29.96	9.84	3.69	-3.23	-16.27	-0.16	2.43
JFC	-6.36	12.77	27.51	4.76	-7.49	-18.34	-35.00	0.12	1.94
KF	10.44	9.79	34.56	16.97	8.81	3.87	-20.32	0.29	2.82
KIF	1.81	10.75	37.02	9.43	-1.29	-6.20	-18.02	0.73	2.89
LAQ	-8.09	10.06	20.11	-1.44	-8.53	-16.72	-26.49	0.38	2.38
LAM	-8.32	10.25	20.39	-1.43	-9.37	-17.06	-27.35	0.56	2.68
LDF	-11.58	7.03	9.59	-7.58	-12.14	-16.26	-27.69	0.35	2.95
MSF	-3.68	7.15	17.47	1.25	-3.08	-8.81	-20.76	-0.07	2.63
MF	-3.13	9.90	40.98	0.81	-5.81	-10.05	-17.71	1.78	6.80
MXE	-4.66	14.86	36.61	5.76	-3.44	-19.76	-28.81	0.36	2.33
MXF	-8.98	10.82	20.99	-3.07	-8.18	-18.34	-31.92	0.40	2.82
ROC	-2.36	11.74	25.83	6.65	-1.51	-11.29	-30.15	-0.22	2.29
SAF	-3.60	10.56	20.51	4.19	-2.45	-13.06	-22.50	0.03	1.98
SGF	2.42	7.60	27.83	7.76	2.63	-3.54	-13.73	0.26	3.06
TWN	1.33	16.24	48.65	13.04	4.81	-11.79	-34.50	-0.15	2.39
EMF	13.52	6.78	35.73	17.96	13.50	7.92	-1.32	0.35	2.60
TC	-0.49	15.54	51.08	1.74	-5.64	-9.65	-21.17	1.56	4.88
TTF	-1.41	16.53	53.44	2.96	-7.25	-11.96	-26.99	1.42	4.39
MEAN	-3.25	10.81	26.73	3.79	-3.59	-11.86	-24.54	0.36	2.82
STD	7.74	2.97	13.42	8.66	8.32	8.34	7.16	0.56	1.13
MAX	14.57	16.53	53.44	23.31	16.82	9.95	-1.32	1.78	6.80
MIN	-20.46	5.96	-6.01	-15.76	-21.41	-25.64	-35.00	-0.74	1.71

Notes:
a) The premia are given in percent, measured by $\ln(P_t/NAV_t)$. UQ and LQ denote he upper and lower quartile, respectively. The median is abbreviated by MED.
b) We also report the skewness (SK) and kurtosis (KU) of the premia distribution.

Table 5.3. Parameter Estimates of the Empirical Premia Process

Fund Name	φ EST	φ STD	χ EST	χ STD	ω EST	ω STD	neg. LogL
AF	3.398***	1.233	-0.025	0.033	0.240***	0.011	2.007
APB	2.020**	0.956	-0.020	0.063	0.283***	0.013	1.831
BZF	3.785***	1.326	-0.067**	0.036	0.299***	0.014	1.792
BZL	4.996***	1.503	-0.062**	0.034	0.383***	0.018	1.559
CH	3.556***	1.214	-0.099***	0.026	0.203***	0.009	2.175
CHN	2.072**	0.964	-0.026	0.062	0.285***	0.013	1.823
CLM	3.128***	1.153	-0.205***	0.021	0.149***	0.007	2.483
MEF	1.935**	0.930	-0.079	0.071	0.301***	0.013	1.770
ETF	1.632**	0.841	-0.090*	0.062	0.223***	0.010	2.067
ISL	1.742**	0.873	-0.077	0.062	0.242***	0.011	1.984
FPF	7.578***	1.913	-0.177***	0.015	0.251***	0.012	2.002
GCH	2.900***	1.121	-0.083**	0.043	0.276***	0.012	1.864
GTF	5.927***	1.633	-0.122***	0.016	0.208***	0.010	2.173
IGF	3.416***	1.311	0.071**	0.039	0.294***	0.013	1.808
IF	4.093***	1.342	0.151***	0.039	0.357***	0.016	1.620
JGF	5.930***	1.660	0.030	0.025	0.329***	0.015	1.719
JFC	2.447***	1.035	-0.076*	0.053	0.288***	0.013	1.819
KF	6.359***	1.726	0.105***	0.025	0.351***	0.016	1.658
KIF	3.532***	1.421	0.034	0.042	0.323***	0.015	1.715
LAQ	3.531***	1.270	-0.092***	0.035	0.274***	0.012	1.877
LAM	2.976***	1.158	-0.092***	0.039	0.256***	0.012	1.938
LDF	9.756***	2.135	-0.113***	0.014	0.304***	0.015	1.831
MSF	6.309***	1.759	-0.043***	0.019	0.261***	0.012	1.951
MF	3.588***	1.434	-0.018	0.038	0.299***	0.014	1.791
MXE	1.573**	0.850	-0.084	0.081	0.276***	0.012	1.851
MXF	2.840***	1.137	-0.105***	0.042	0.266***	0.012	1.900
ROC	2.770***	1.110	-0.034	0.046	0.282***	0.013	1.841
SAF	2.630***	1.090	-0.049	0.043	0.249***	0.011	1.962
SGF	7.092***	1.797	0.025*	0.018	0.284***	0.013	1.876
TWN	1.857**	0.916	-0.001	0.079	0.325***	0.015	1.692
EMF	9.436***	2.143	0.133***	0.014	0.295***	0.014	1.859
TC	1.570**	0.877	0.038	0.088	0.298***	0.013	1.776
TTF	1.408**	0.868	0.046	0.106	0.313***	0.014	1.724
MEAN	3.872	1.294	-0.037	0.043	0.281	0.013	1.871
STD	2.245	0.374	0.082	0.023	0.046	0.002	0.178
MAX	9.756	2.143	0.151	0.106	0.383	0.018	2.483
MIN	1.408	0.841	-0.205	0.014	0.149	0.007	1.559

Note:
Statistically significant parameter estimates at the 1%-, 5%- and 10%-levels are denoted by ***, **, and *, respectively.

The range is from plus 15 percent to discounts of 21 percent which is close to the descriptive results for the mean values of table 5.2. When comparing the estimation results with the three closed-end funds shown in figures 4.1, 4.3 and 4.2, we can confirm the high mean-reversion tendency of 5.9 for the GT Global Eastern Europe (GTF). We can also infer the high volatility for the Indonesia Fund (IF) from the figure. The long-run means for the closed-end funds IF and GTF are estimated in line with values of plus 15 percent and minus 12 percent.

5.2 Implemented Model

In order to implement the theoretical valuation model for closed-end fund shares on capital market data we need to choose an appropriate empirical model which comes close to the properties of the general model of equation (4.8). Examining the available data on closed-end funds we notice that there exist virtually no closed-end funds that feature automatic windup dates.[6] Therefore, we deal with infinite time horizons compared to an explicit expiration date T as generally given in applications to price contingent claims. Thus, for the purpose of the empirical implementation of our valuation model we need to ensure the boundedness of the price stated in equation (4.8) for the limit of T reaching infinity. Examining the equation for the limits $T \to \infty$ yields:

$$\begin{aligned} Y(X, \pi, t; \psi_M) = X_t - \frac{1}{\kappa}(\pi_t - \theta) - \frac{1}{\kappa^2}\sigma_\pi \left(\lambda_\pi - \rho\sigma_X + \frac{3}{4\kappa}\sigma_\pi\right) \\ - \lim_{T\to\infty}\left(\theta - \frac{1}{\kappa}\sigma_\pi\lambda_\pi + \frac{1}{\kappa}\rho\sigma_X\sigma_\pi - \frac{\sigma_\pi^2}{2\kappa^2}\right)(T-t). \end{aligned}$$

Thus, we actually reach a steady state except for the last expression which remains from the first term of the function $C(t)$. Therefore, we constrain our model to the case where the long-run mean of the continuous premium is given by

$$\theta = \frac{1}{\kappa}\sigma_\pi\lambda_\pi - \frac{1}{\kappa}\rho\sigma_X\sigma_\pi + \frac{\sigma_\pi^2}{2\kappa^2}, \tag{5.2}$$

which reduces our parameter space by one parameter. Using equation (5.2) we obtain a functional mapping of the two state variables X_t and π_t to the value

[6] There are very rare exceptions which show this feature and no closed-end funds included in our sample.

of the observable asset $Y(X, \pi, t; \psi_M)$. The pricing relationship for the implementation of the closed-end fund market prices according to the valuation model is given by

$$P(X, \pi, t; \psi_M) = e^{-\frac{1}{\kappa}(\pi_t - \theta) - \frac{1}{\kappa^2}\sigma_\pi\left(\lambda_\pi - \rho\sigma_X + \frac{3}{4\kappa}\sigma_\pi\right)} NAV_t, \tag{5.3}$$

where $P(X, \pi, t; \psi_M)$ depends on the model parameters $\psi_M = \{\sigma_X, \kappa, \sigma_\pi, \rho, \lambda_\pi\}$. In further examining the valuation formula of equation (5.3) we are finally able to show how the second factor we choose in the valuation model is related to the empirical premium as defined in equation (4.1). By rewriting the valuation formula in terms of logarithms we obtain the connecting formula

$$PREM_t = -\frac{1}{\kappa}\pi_t + const \tag{5.4}$$

for the two premia, with a constant term denoted by $const = f(\psi_M)$. From this linear relationship between the empirical and the modeled continuous premia we expect to observe positive dynamic premia on closed-end funds with corresponding empirical discounts, and vice versa, in the empirical analysis.

5.3 State Space Form

In this section we present the empirical adaptation of the theoretical pricing model as stated in equation (5.3). In order to empirically estimate the model parameters, we present the dynamic model in a state space form.[7] Therefore, we first specify the measurement and the transition equation, to then be able to use a Kalman filter algorithm to perform a maximum likelihood estimation of the parameters and to obtain a time-series of the continuous premia. For the state variables we choose $\boldsymbol{\xi}_t = [X_t, \pi_t - \theta]'$ which are assumed to follow the *transition equation*

$$\boldsymbol{\xi}_t = \mathbf{c} + \Phi\boldsymbol{\xi}_{t-\Delta t} + \boldsymbol{\eta}_t \tag{5.5}$$

for a time interval of Δt. In order to specify the corresponding state space form to equations (4.2) and (4.3) with the *exact discrete time equivalents*, i.e. the discrete approximation yields exactly the same distributions as in the continuous case, we need to match the following moments.[8]

---

[7] For more details on state space modeling compare section 2.1 and see, for example, Aoki (1990).

[8] See, for example, Bergstrom (1984).

Remark 5.1 (Moments of the State Variables) *For the first factor* $\boldsymbol{\xi}_{1,t} = X_t$ *we use equation (4.2) which can be integrated over the time period* $\Delta t = [t-\Delta t, t]$ *to yield the almost explicit solution*

$$X_t = X_{t-\Delta t} + \left(\mu - \frac{1}{2}\sigma_X^2\right)\Delta t + \sigma_X \int_{t-\Delta t}^{t} dW_{X,t}.$$

Based on equation (4.3) for the second state variable, $\boldsymbol{\xi}_{2,t} = \pi_t - \theta$*, we consider the differential of the function* $F(\pi_t, t) = e^{\kappa t}(\pi_t - \theta)$ *by applying Itô's formula*

$$de^{\kappa t}(\pi_t - \theta) = \kappa e^{\kappa t}(\pi_t - \theta)\,dt + e^{\kappa t} d\pi_t = \sigma_\pi e^{\kappa t} dW_{\pi,t}$$

which we also integrate to find

$$(\pi_t - \theta) = e^{-\kappa \Delta t}(\pi_{t-\Delta t} - \theta) + e^{-\kappa t}\sigma_\pi \int_{t-\Delta t}^{t} e^{\kappa\varsigma} dW_{\pi,\varsigma}$$

besides solving for the Itô integral. Since the remaining integrals are martingales we are able to calculate the conditional moments to be[9]

$$\mathbb{E}\left[\boldsymbol{\xi}_t | \mathcal{F}_{t-\Delta t}\right] = \begin{bmatrix} \left(\mu - \frac{1}{2}\sigma_X^2\right)\Delta t \\ 0 \end{bmatrix} + \begin{bmatrix} 1 & 0 \\ 0 & e^{-\kappa\Delta t} \end{bmatrix} \boldsymbol{\xi}_{t-\Delta t}, \text{ and} \tag{5.6}$$

$$\mathbb{V}ar\left[\boldsymbol{\xi}_t | \mathcal{F}_{t-\Delta t}\right] = \begin{bmatrix} \sigma_X^2 & \rho\sigma_X\sigma_\pi \\ \rho\sigma_X\sigma_\pi & \sigma_\pi^2 \end{bmatrix} \Delta t. \tag{5.7}$$

For the variance term we use an Euler approximation.

Thereby, the appropriate matrix and moment specifications of the transition equation (5.5) are given by

$$\mathbf{c} = \left[\left(\mu - \tfrac{1}{2}\sigma_X^2\right)\Delta t, 0\right]', \; \Phi = \begin{bmatrix} 1 & 0 \\ 0 & e^{-\kappa\Delta t} \end{bmatrix}, \text{ and}$$

$$\mathbb{E}[\boldsymbol{\eta}_t] = [0,0]', \qquad \mathbb{E}[\boldsymbol{\eta}_t \boldsymbol{\eta}_s'] = \begin{bmatrix} \sigma_X^2 & \rho\sigma_X\sigma_\pi \\ \rho\sigma_X\sigma_\pi & \sigma_\pi^2 \end{bmatrix} \Delta t \cdot \delta_{t,s}$$

using a discretisation length of $\Delta t = 1/52$ for weekly data.[10]

---

[9] Using Itô's isometry and the fact that the expectation of an Itô integral is zero. See, for example, Oksendal (2000).

[10] The function $\delta_{i,j}$ stands for the Kronecker symbol which takes value 1 if $i = j$, and 0 else.

With the two observable prices on closed-end funds, the net asset values and their market values defined by $\mathbf{Z}_t = [X_t, Y_t]'$, we obtain the *measurement equation*

$$\mathbf{Z}_t = \mathbf{a} + \mathbf{B}\boldsymbol{\xi}_t + \varepsilon_t. \tag{5.8}$$

Equation (5.8) serves as a link between the latent state variables $\boldsymbol{\xi}_t$ and the observable net asset values NAV_t and the market prices P_t for the shares. These prices are assumed to be only measurable with a noise $\varepsilon_t = [\varepsilon_{X,t}, \varepsilon_{Y,t}]'$.[11] For the error terms we further assume $\mathbb{E}[\varepsilon_t] = \mathbf{0}$, and $\mathbb{E}[\varepsilon_t \varepsilon_s'] = \mathbf{H} \cdot \delta_{t,s}$.[12] In order to obtain our empirical model of equation (5.3) developed in the previous section, we specify the distinct coefficients

$$\mathbf{a} = \left[0, -\frac{1}{\kappa^2}\sigma_\pi \left(\lambda_\pi - \rho\sigma_X + \frac{3}{4\kappa}\sigma_\pi\right)\right]',$$
$$\mathbf{B} = \begin{bmatrix} 1 & 0 \\ 1 & -1/\kappa \end{bmatrix}, \text{ and } \mathbf{H} = \begin{bmatrix} h_X^2 & 0 \\ 0 & h_Y^2 \end{bmatrix}.$$

Combining the transition and the measurement equations, the full state space model depends on the time homogeneous parameter vector $\psi_{SSM} = \{\kappa, \lambda_\pi, \sigma_\pi, h_Y, \mu, \sigma_X, h_X, \rho\}$. These parameters are estimated in a statistical inference using a recursion algorithm based on linear *Kalman filtering* as described in chapter 4. With the specified state space form of equations (5.5) and (5.8), we are able to perform a maximum likelihood estimation of the parameters

$$\hat{\psi}_{SSM} = \underset{\psi_{SSM} \in \Psi}{\arg\max} L\left(\mathbf{Z}_T, \mathbf{Z}_{T-1}, \dots, \mathbf{Z}_1; \psi_{SSM}\right). \tag{5.9}$$

Additional to the parameter estimation, the Kalman filter enables us to extract a time-series of the unobservable state variables $\boldsymbol{\xi}_t$ from the data. In the following section we present a closed-end fund analysis based on the examined sample of closed-end fund data.

5.4 Closed-End Fund Analysis

In this empirical analysis we examine the characteristics of our pricing model by estimating its model parameters and the resulting time-series for the dynamic premia. The inference is based on the maximum likelihood estimator of equation (5.9) using the defined Kalman filter setup of section 5.3.

[11] The small errors ε_t are introduced to capture imperfections in observed price quotations.

[12] Further, we assume independent error terms ε_t and η_t.

The results on the estimated values of the parameters contained in ψ_{SSM} are shown in table 5.4 and are continued in table 5.5. We first analyze the parameters corresponding to the second state variable, i.e. the dynamic premium π_t. The coefficient of mean-reversion on the continuous premia is estimated by 1.6 on average across all funds. Interpreting these parameter estimates of κ as half-life of the premia[13], we infer a range of 0.1 to 1.6 years for the different closed-end funds with a mean value of 0.4 years. These numbers indicate high fluctuations of the dynamic premia around their long-run means θ. This fact is emphasized in that closed-end funds with high κ values also show a high standard deviation of the premium. For the volatility on the dynamic premia σ_π we average around 0.3 with outliers that go up to 1.0. Further, the estimated values for the market price of risk λ_π show no distinct pattern in that it varies from -3 to +5 with a mean close to zero of 0.3. Finally, the volatility of the error terms h_Y is estimated significantly around the value of 1.8% on average.

The outcome on the remaining parameter estimates are reported in table 5.5. First, we show the results on the drift coefficient μ for which we obtain an average value of 5 percent with a standard deviation of 12 %; the estimates, however, are not statistically significant in general. The estimates of the net asset value volatility σ_X closely resemble the values obtained empirically from the data (compare table 5.1). The differences take a maximum of 4.6 percentage points with an average of 1.1%. Further, the estimated model error h_Y of 0.005 across all funds translates to average pricing deviations of 0.54% with the observable prices on the net asset values.[14] Lastly, the values for the correlation coefficient ρ primarily show positive values up to +0.5 with a mean of +0.13, i.e. changes in the net asset value and the dynamic premium are related.

Further, the results on the parameter estimates of the volatility terms σ_π and σ_X of the used state variables show that the risk of altering premia with an average of 30.4% is about 1.3 times higher than the risk involved with the net asset value showing a value of 23.7%. This, however, is a result directly related to our valuation model; the empirical standard deviations on the closed-end fund sample show a mean of the empirical premia of only 10.8% which is lower than both, the empirical as well as the estimated net asset value volatility (compare tables 5.1 and 5.2).

[13] The half-life of the dynamic premium is given by $\ln(2)/\kappa$. This can be deferred from equation (5.6) solving for $\tau = t - s$.

[14] See, for example, Aitchison and Brown (1957, ch. 2.2) for the log-normal transformation.

Table 5.4. Parameter Estimates of ψ_{SSM}

Fund	κ		λ_π		σ_π		h_Y	
Name	EST	STD	EST	STD	EST	STD	EST	STD
AF	0.976***	0.011	0.197	0.568	0.143***	0.001	0.018***	0.000
APB	0.773***	0.000	0.024	0.144	0.125***	0.004	0.022***	0.001
BZF	0.810***	0.003	0.519	0.340	0.113***	0.001	0.019***	0.000
BZL	1.288***	0.005	0.499	0.348	0.248***	0.000	0.027***	0.000
CH	3.091***	0.000	1.919***	0.519	0.453***	0.001	0.014***	0.000
CHN	1.118***	0.119	0.375	0.393	0.219***	0.001	0.017***	0.000
CLM	1.017***	0.000	2.687***	0.703	0.078***	0.017	0.011***	0.000
MEF	0.891***	0.001	0.247	0.330	0.182***	0.000	0.021***	0.000
ETF	0.423***	0.002	0.230	0.237	0.049***	0.000	0.019***	0.000
ISL	1.337***	0.001	0.739**	0.356	0.247***	0.000	0.015***	0.000
FPF	4.630***	0.013	5.045***	0.437	0.778***	0.003	0.016***	0.000
GCH	1.194***	0.004	0.720**	0.366	0.194***	0.000	0.020***	0.000
GTF	2.119***	0.003	2.439***	0.384	0.242***	0.001	0.014***	0.000
IGF	1.548***	0.000	-0.185	0.397	0.391***	0.001	0.010***	0.000
IF	2.494***	0.006	-1.369***	0.399	0.637***	0.004	0.020***	0.000
JGF	2.034***	0.002	-0.401	0.416	0.379***	0.002	0.021***	0.000
JFC	1.263***	0.005	0.508	0.407	0.241***	0.001	0.019***	0.000
KF	1.220***	0.000	-0.886***	0.356	0.383***	0.007	0.010***	0.000
KIF	1.840***	0.006	-0.264	0.346	0.456***	0.000	0.013***	0.000
LAQ	1.047***	0.002	-0.082	0.358	0.165***	0.000	0.021***	0.000
LAM	1.056***	0.058	-0.066	0.292	0.224***	0.017	0.010***	0.002
LDF	0.813***	0.002	0.089	0.358	0.162***	0.001	0.020***	0.000
MSF	1.130***	0.003	-0.076	0.344	0.193***	0.000	0.019***	0.000
MF	2.070***	0.001	-0.320	0.370	0.459***	0.004	0.018***	0.000
MXE	0.768***	0.002	0.060	0.251	0.145***	0.002	0.019***	0.000
MXF	1.037***	0.009	0.498**	0.224	0.175***	0.001	0.042***	0.000
ROC	1.785***	0.002	-0.033	0.170	0.362***	0.001	0.019***	0.000
SAF	1.265***	0.002	0.214	0.478	0.200***	0.007	0.018***	0.000
SGF	4.513***	0.000	-0.575	0.451	1.010***	0.004	0.015***	0.000
TWN	0.822***	0.000	-0.021	0.036	0.189***	0.001	0.019***	0.000
EMF	4.298***	0.021	-3.177***	0.436	0.802***	0.002	0.020***	0.000
TC	0.845***	0.002	-0.230	0.345	0.157***	0.000	0.021***	0.000
TTF	0.911***	0.000	-0.048	0.281	0.221***	0.002	0.018***	0.000
MEAN	1.589	0.009	0.281	0.359	0.304	0.003	0.018	0.000
STD	1.085	0.022	1.319	0.122	0.223	0.004	0.006	0.000
MAX	4.630	0.119	5.045	0.703	1.010	0.017	0.042	0.002
MIN	0.423	0.000	-3.177	0.036	0.049	0.000	0.010	0.000

Note:
Statistically significant parameter estimates at the 1%-, 5%- and 10%-levels are denoted by ***, **, and *, respectively.

Table 5.5. Parameter Estimates of ψ_{SSM} (continued)

Fund	μ		σ_X		h_X		ρ		neg.
Name	EST	STD	EST	STD	EST	STD	EST	STD	LogL
AF	0.148	0.099	0.220***	0.004	0.002***	0.000	-0.119***	0.000	1061.88
APB	0.056	0.091	0.207***	0.000	0.000***	0.000	0.024***	0.000	1039.75
BZF	0.229	0.145	0.328***	0.001	0.018***	0.000	0.153***	0.000	898.02
BZL	0.237	0.160	0.370***	0.002	0.017***	0.000	0.278***	0.000	808.01
CH	0.143*	0.078	0.175***	0.000	0.000***	0.000	-0.049***	0.000	1161.79
CHN	0.058	0.084	0.194***	0.000	0.009***	0.000	0.008***	0.001	1026.79
CLM	0.108**	0.046	0.100***	0.041	0.005***	0.000	-0.011***	0.002	1378.02
MEF	0.065	0.141	0.327***	0.000	0.002***	0.000	0.304***	0.000	901.30
ETF	0.149**	0.072	0.160***	0.000	0.000***	0.000	-0.138***	0.001	1167.38
ISL	0.062	0.083	0.180***	0.000	0.003***	0.000	0.153***	0.000	1106.93
FPF	0.000	0.052	0.223***	0.001	0.000***	0.000	0.255***	0.001	1062.24
GCH	0.102	0.098	0.231***	0.001	0.008***	0.000	0.079***	0.000	1004.23
GTF	0.152**	0.069	0.153***	0.000	0.009***	0.000	-0.113***	0.000	1176.75
IGF	0.015	0.048	0.228***	0.001	0.009***	0.000	0.237***	0.000	989.07
IF	-0.113	0.121	0.282***	0.001	0.010***	0.000	0.304***	0.002	892.83
JGF	-0.118	0.116	0.267***	0.003	0.012***	0.000	0.250***	0.001	935.04
JFC	0.013	0.594	0.212***	0.000	0.008***	0.000	0.098***	0.000	1012.46
KF	-0.133	0.145	0.325***	0.002	0.003***	0.000	0.301***	0.002	865.20
KIF	-0.187	0.134	0.307***	0.001	0.015***	0.000	0.236***	0.000	884.25
LAQ	0.149*	0.089	0.208***	0.000	0.007***	0.000	-0.252***	0.000	1037.28
LAM	0.109	0.087	0.199***	0.010	0.007***	0.002	-0.160***	0.000	1048.51
LDF	0.145	0.120	0.282***	0.001	0.004***	0.000	0.284***	0.001	946.94
MSF	0.071	0.074	0.163***	0.000	0.000***	0.000	0.048***	0.000	1115.73
MF	-0.155	0.120	0.288***	0.000	0.000***	0.000	0.501***	0.004	959.44
MXE	0.214*	0.126	0.275***	0.001	0.001***	0.000	0.104***	0.000	963.84
MXF	0.030	0.159	0.355***	0.003	0.004***	0.000	0.246***	0.000	869.54
ROC	0.067	0.079	0.226***	0.000	0.004***	0.000	-0.115***	0.000	1007.59
SAF	0.035	0.104	0.191***	0.000	0.001***	0.000	0.087***	0.003	1090.93
SGF	0.015	0.078	0.177***	0.000	0.003***	0.000	0.116***	0.000	1069.41
TWN	0.133	0.105	0.233***	0.001	0.011***	0.000	0.158***	0.000	946.06
EMF	0.154**	0.072	0.164***	0.002	0.000***	0.000	0.289***	0.000	1106.40
TC	-0.176	0.125	0.275***	0.001	0.006***	0.000	0.271***	0.000	951.12
TTF	-0.137	0.131	0.311***	0.001	0.001***	0.000	0.416***	0.000	911.17
MEAN	0.050	0.116	0.237	0.002	0.005	0.000	0.129	0.001	1012.00
STD	0.120	0.091	0.066	0.007	0.005	0.000	0.180	0.001	115.54
MAX	0.237	0.594	0.370	0.041	0.018	0.002	0.501	0.004	1378.02
MIN	-0.187	0.046	0.100	0.000	0.000	0.000	-0.252	0.000	808.01

Note:
Statistically significant parameter estimates at the 1%-, 5%- and 10%-levels are denoted by ***, **, and *, respectively.

Table 5.6. Values of θ and Statistic of Dynamic Premia π_t

Fund		$\pi_t^{b)}$								
Name	$\theta^{a)}$	MEAN	STD	MAX	UQ	MED	LQ	MIN	SK[c)]	KU[c)]
AF	4.335	-0.20	8.60	19.30	6.83	-2.33	-7.17	-15.97	0.34	2.15
APB	1.624	-0.45	10.24	16.68	9.89	-1.77	-9.19	-20.96	0.01	1.67
BZF	7.489	4.15	8.13	20.71	12.91	2.50	-2.65	-10.40	0.33	1.86
BZL	9.468	6.51	14.58	37.56	21.50	5.53	-5.48	-26.77	-0.05	2.05
CH	29.341	26.13	23.28	70.94	43.26	29.07	12.91	-31.60	-0.48	2.42
CHN	9.233	0.54	14.87	30.41	12.41	-1.49	-11.51	-31.02	0.27	1.95
CLM	20.841	20.67	5.85	30.11	25.83	21.61	15.81	7.51	-0.33	1.91
MEF	5.114	3.56	12.98	23.45	15.96	2.71	-2.94	-29.03	-0.55	2.64
ETF	3.601	2.67	5.02	9.56	7.49	2.95	-1.12	-8.84	-0.47	2.03
ISL	14.874	8.26	16.71	36.87	20.94	14.03	-1.66	-30.75	-0.74	2.42
FPF	85.265	83.45	27.75	135.50	101.18	87.94	75.33	-26.02	-1.58	6.15
GCH	12.752	8.35	13.15	34.58	21.72	8.44	-3.31	-17.61	0.02	1.78
GTF	28.684	25.65	12.06	45.39	34.32	28.93	18.88	-7.04	-0.82	2.83
IGF	-2.854	-14.10	16.09	25.88	-2.50	-12.60	-23.52	-68.83	-0.37	3.40
IF	-33.910	-37.88	29.96	30.00	-28.14	-43.34	-60.28	-94.02	0.81	2.82
JGF	-6.969	-6.49	18.08	29.98	5.82	-9.40	-20.19	-46.33	0.32	2.30
JFC	11.124	7.13	15.75	39.80	22.58	8.32	-6.70	-31.14	-0.10	1.85
KF	-25.960	-15.21	11.84	21.17	-7.27	-13.09	-23.52	-44.24	-0.30	2.77
KIF	-5.262	-4.84	19.24	29.31	10.20	1.02	-18.54	-65.34	-0.71	2.69
LAQ	0.779	7.72	10.18	27.64	16.38	7.68	1.25	-16.25	-0.32	2.26
LAM	1.522	7.64	10.73	27.50	16.69	8.53	0.47	-22.35	-0.55	2.68
LDF	2.164	8.40	5.45	19.55	11.82	8.91	4.76	-6.66	-0.31	2.82
MSF	0.029	3.37	7.69	21.03	8.86	2.72	-1.70	-16.41	0.12	2.51
MF	-7.817	5.18	19.95	32.59	18.63	10.60	-2.25	-76.84	-1.81	6.79
MXE	2.364	2.65	11.25	20.36	14.41	1.73	-5.20	-26.62	-0.34	2.26
MXF	8.367	8.49	10.60	29.24	18.03	7.31	2.88	-19.34	-0.36	2.72
ROC	1.917	3.12	20.42	51.26	18.29	1.42	-12.36	-41.07	0.26	2.24
SAF	4.356	3.90	12.96	25.66	15.91	2.45	-5.56	-23.49	0.01	1.94
SGF	-10.817	-12.15	33.02	56.99	13.36	-14.12	-34.29	-113.76	-0.22	3.01
TWN	1.319	-2.42	13.14	25.98	8.09	-5.24	-12.22	-35.95	0.18	2.31
EMF	-58.447	-59.01	26.67	-8.21	-35.08	-58.70	-77.13	-139.25	-0.30	2.36
TC	-3.932	-0.42	12.90	16.08	6.96	3.68	-1.98	-42.19	-1.58	4.81
TTF	-1.375	-0.17	14.96	22.96	9.15	5.37	-4.14	-49.40	-1.43	4.38
MEAN	3.310	2.85	14.97	32.00	14.44	3.37	-6.74	-37.21	-0.34	2.75
STD	21.975	21.38	6.97	23.24	21.33	22.58	23.95	31.53	0.60	1.17
MAX	85.265	83.45	33.02	135.50	101.18	87.94	75.33	7.51	0.81	6.79
MIN	-58.447	-59.01	5.02	-8.21	-35.08	-58.70	-77.13	-139.25	-1.81	1.67

Notes:
a) The long-run means of the dynamic premia (in percent) are given by equation (5.2).
b) In the statistics for the dynamic premia (in percent), UQ and LQ denote the upper and lower quartile respectively; the median is abbreviated by MED.
c) For each closed-end fund we also report the skewness (SK) and kurtosis (KU) of the dynamic premium distribution.

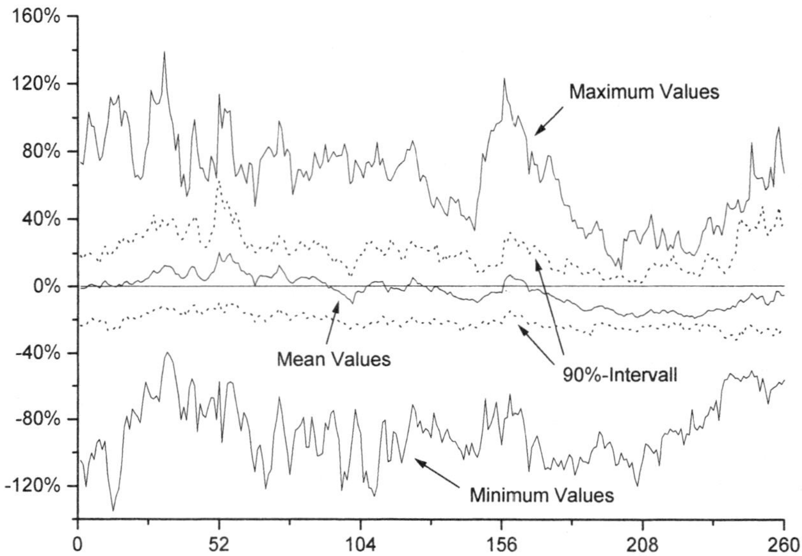

Fig. 5.1. Time Series of (Inverted) Dynamic Premia

Taking a final look at the significance levels of the estimated parameters, we find that they show highly significant values for the parameters throughout the various funds, except for the estimates of the drift parameters λ_π and μ which is a typical phenomenon in financial econometrics. However, the estimates for the market price of risk λ_π show significant values for selected funds, such as CLM, FPF, KF, and EMF which trade at high discounts and premia (compare table 5.2).

The results for the time-series of the dynamic premia for the different closed-end funds are reported in table 5.6 along with their theoretical long-run mean values. The sample means (MEAN) take values close to the long-run means θ. The differences are below eight percent, except for the funds CHN, IGF, KF, and MF, where the MF fund shows the maximum difference of 13%. The means themselves exhibit a wide range of -53 to +83 percent. But only 13 out of the 33 closed-end funds show a higher empirical than theoretical mean value. Examining the sign of the dynamic premia indicated by our theoretical result of equation (5.4), we find a negative relation between the signs of the dynamic and the empirical premia. Our observations coincide with this theoretical implication except for the minor deviations of the AF, APB, TC, and TTF funds. However, the sign needs to be put into perspective when incorporating the reported standard deviations. Looking at the realized

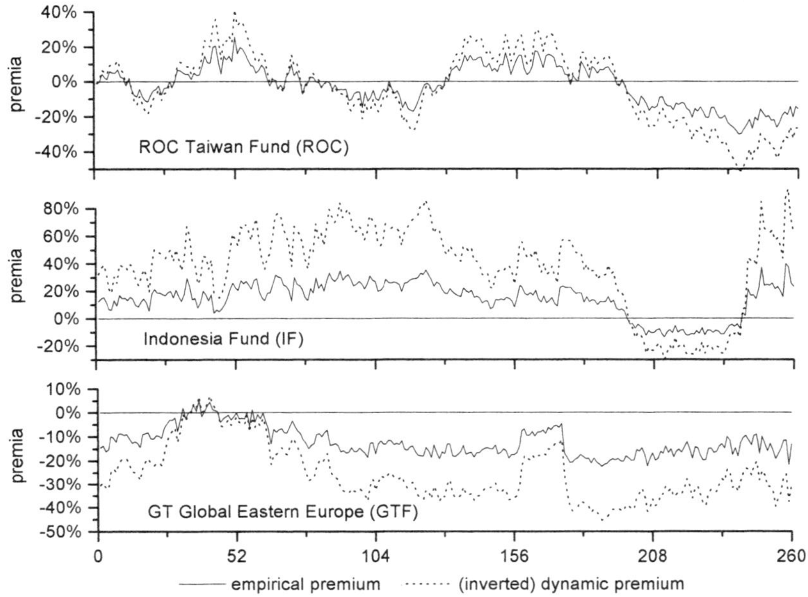

Fig. 5.2. Comparison of Empirical and Dynamic Premia

dynamic premia, we find high fluctuations as indicated by the estimated values for the speed of mean-reversion and the volatility from above. With 135.5 percent FPF marks the maximum and EMF shows the minimum of minus 139.3 percent.

In figure 5.1 we graph the evolution of the dynamic premia with inverted sign over the sample period, i.e. we visualize the numerical results of table 5.6. The figure contains the maximum and minimum values across all funds, along with the mean values and the 90 percent interval. In general we see a higher fluctuation in the positive premia from the maximum values as well as the confidence interval. The negative premia seem to be estimated in a much closer range. Overall, the visual impression resembles the results for the empirical premia as shown in figure 4.4 on page 52 except for a scaling factor. This fact also holds when we look at the three selected funds from section 4.1 which we graph in figure 5.2. In each of the graphs we plot the time-series of the empirical as well as the (inverted) dynamic premia for the three closed-end funds ROC, IF, and GTF. We see that for the funds IF and GTF the inverted dynamic premia emphasizes the positive and negative values.

6 Implications for Investment Strategies

In this chapter we further examine potential implications of the pricing model for investment strategies. Based on the empirical findings presented in the previous chapter we infer insights into two suggested applications relevant for closed-end fund investors. *First*, we address the investors' fundamental question of the forecasting power of the pricing model, i.e. we investigate the ability of the pricing model to predict the market prices of the closed-end funds based on current information. The *second* question we raise is to test on how much information content lies in the estimated values of the closed-end fund premia. For this purpose we study trading strategies that exploit the observable differences between the dynamic premia and their long-run equilibria.

6.1 Testing the Forecasting Power

6.1.1 Setup of Forecasting Study

The study to test the forecasting quality of the pricing model is setup as follows and is illustrated in figure 6.1. From the available five years of the closed-end fund sample we choose the first four years, which contain 208 weekly observations, to estimate the parameter vector ψ_{SSM} for the various funds. The estimation results are close to the reported parameters for the whole sample period (see tables 5.4 and 5.5) and thus are not reported separately. The last 52 weeks of the sample are used to test whether forecasts up to four weeks beginning at the X-th week can generate meaningful price predictions of the closed-end fund market shares in an out of sample setting. In order to get current values for the state variables at initiation of the forecast we filter the data based on the estimated four year parameters for the period of the first week in the sample to the X-th week.

The finite sample forecasts are obtained from the same state space form as used with the Kalman filter. The expected values for the s-period-ahead predictions of the state variables and the observable prices are the conditional expectations

$$\boldsymbol{\xi}_{t+s\cdot\Delta t|t} = \Phi^s \boldsymbol{\xi}_{t|t}, \text{ and } \mathbf{Y}_{t+s\cdot\Delta t|t} = \mathbf{a} + \mathbf{B}\boldsymbol{\xi}_{t+s\cdot\Delta t|t},$$

which can be deferred from the law of iterated projections.[1]

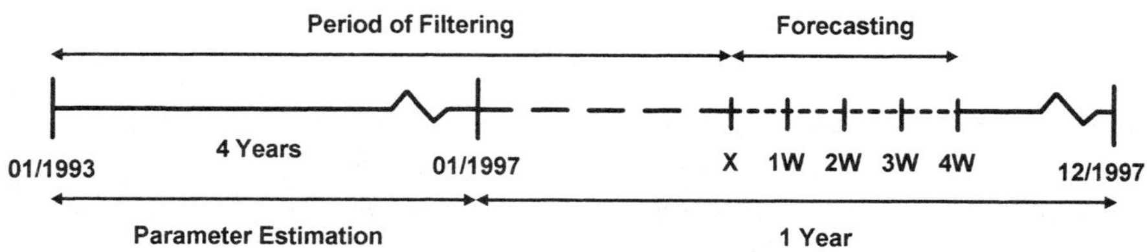

Fig. 6.1. Setup of Forecasting Study

Further, we increment the initial week successively by one week starting from the 208th week and ending with week 256. This forecasting procedure creates 49 sets of four week forecasts on the various closed-end fund market prices.

In order to evaluate the quality of the forecast results we additionally implement a reference model. For this purpose we choose a most general Gaussian $ARMA(1,1)$-model for log-share prices Y_t given by

$$Y_t - m = a\left(Y_{t-\Delta t} - m\right) + \varepsilon_t + b\varepsilon_{t-\Delta t}$$

with $\Delta t = 1/52$. This process specification of the closed-end funds market prices translates with the matrix specifications

$$\mathbf{c} = \mathbf{0}, \;\; \Phi = \begin{bmatrix} a & 0 \\ 1 & 0 \end{bmatrix}, \boldsymbol{\eta}_t = \begin{bmatrix} \varepsilon_t \\ 0 \end{bmatrix}, \mathbb{E}\left[\boldsymbol{\eta}_t \boldsymbol{\eta}_s'\right] = \begin{bmatrix} \sigma^2 & 0 \\ 0 & 0 \end{bmatrix} \cdot \delta_{t,s},$$
$$\mathbf{a} = m, \mathbf{B} = [1, b], \;\; \text{and} \qquad \mathbb{E}\left[\varepsilon_t \varepsilon_s\right] = \delta_{t,s}$$

into a two-factor state space representation comparable to that of equations (5.5) and (5.8).[2] Thereby, the setup for parameter estimation and forecasting is analogous to the one for the original pricing model.[3]

[1] See, for example, Harvey (1989).
[2] See Hamilton (1994b, p. 387).
[3] The maximum likelihood results are reliably estimated and not reported.

6.1.2 Evidence on Forecasting Quality

The results on the forecasting study are reported in table 6.1. The table contains the descriptive statistics stated in prices for the valuation model as well as their differences to the reference model. Generally, to calculate the prediction error we subtract the forecast results of the implemented models from the realized closed-end fund prices on the market. The five selected statistics are calculated separately for the 33 funds in the sample and described by their average values (MEAN), their standard deviation (STD), the maximum (MAX), and the minimum (MIN) for all closed-end funds in the sample.

The first five rows show the mean results over the various funds. The average forecasting error of our pricing model in absolute terms (mean absolute deviations, MAD) is 59 cents for one week ahead predictions and continuously increases to \$1.29 for the one month period; the corresponding root mean square errors (RMSE) are slightly higher with values from 82 cents for one week up to \$1.76 for the longest period. Compared to the reference model the predictions are better throughout the different forecasting periods by 3 to 8 cents; for the RMSE the results are slightly better for the first week but weaken for longer forecasting periods. Looking at the mean maximum and minimum price deviations we find our model to perform better for all four weeks ranging from 14 up to 69 cents.

The next rows contain information on the variability (STD) of the five statistics over the different funds. The goodness of fit measured by the MAD (RMSE) ranges within a one-sigma-interval from 36 to 83 cents (49 cents to \$1.14) for the one week ahead predictions. For one month forecasts the interval widens to 85 cents and \$1.74 (\$1.10 and \$2.43). Compared to the reference model the results on the variability are better for the MAD measure, but higher in the case of RMSE.

In the last part of table 6.1 we document the maximum and minimum values of the statistics. We find the realized maximum MAD to lie between \$1.13 and \$2.37 which is throughout lower than the results for the reference model. The highest individual prediction errors take values from \$4.26 up to \$5.32 for one month. The corresponding minimum realizations of the forecasting errors are found to range from -\$5.08 to -\$12.04. This indicates that the pricing model tends to overestimate the forthcoming price movements and provides evidence that the forecasting power is better in times of rising market prices. Negative price jumps are only captured with higher forecast errors which, however, are lower than those predicted by the reference model.

Table 6.1. Results on the Forecasting Study in Terms of Market Prices

[Results and Differences in $]		Model Results 1W	2W	3W	4W	Difference to ARMA(1,1)[d] 1W	2W	3W	4W
MEAN	SSE[a]	37.553	77.659	126.912	173.589	-0.410	0.401	4.850	9.074
	MAD[b]	0.595	0.862	1.100	1.293	-0.029	-0.045	-0.059	-0.078
	RMSE[b]	0.815	1.174	1.504	1.764	-0.007	-0.001	0.021	0.037
	RES_{MAX}[c]	1.774	2.082	2.317	2.256	-0.137	-0.298	-0.361	-0.574
	RES_{MIN}[c]	-2.450	-3.657	-4.498	-5.248	-0.179	-0.275	-0.538	-0.692
STD	SSE[a]	29.968	62.577	101.710	135.607	-0.035	1.621	7.732	12.954
	MAD[b]	0.231	0.324	0.397	0.446	-0.019	-0.026	-0.028	-0.039
	RMSE[b]	0.324	0.461	0.581	0.667	0.004	0.011	0.032	0.045
	RES_{MAX}[c]	0.939	1.208	1.222	1.214	0.006	0.013	0.023	-0.047
	RES_{MIN}[c]	1.148	1.882	2.208	2.460	0.163	0.286	0.409	0.503
MAX	SSE[a]	118.122	290.061	488.989	644.715	-1.973	9.839	36.883	65.832
	MAD[b]	1.128	1.670	2.078	2.375	-0.122	-0.130	-0.163	-0.195
	RMSE[b]	1.553	2.433	3.159	3.627	-0.013	0.042	0.121	0.190
	RES_{MAX}[c]	4.256	6.240	5.494	5.315	-0.041	0.695	-0.028	-0.739
	RES_{MIN}[c]	-0.919	-1.308	-1.654	-1.720	0.039	0.130	-0.329	-0.483
MIN	SSE[a]	5.193	10.038	15.472	20.068	-1.302	-2.788	-4.684	-6.935
	MAD[b]	0.233	0.351	0.460	0.535	-0.038	-0.069	-0.096	-0.121
	RMSE[b]	0.326	0.453	0.562	0.640	-0.039	-0.059	-0.079	-0.102
	RES_{MAX}[c]	0.672	0.379	0.462	0.479	0.099	-0.203	-0.341	-0.452
	RES_{MIN}[c]	-5.076	-8.831	-10.683	-12.038	-0.241	-0.902	-1.283	-1.661

Notes:
a) We first report the sum of squared errors (SSE).
b) The two measures to evaluate the goodness of fit are the mean absolute deviation (MAD) and the root mean square error (RMSE).
c) We further show the results on the maximum and the minimum of the prediction errors (RES_{MAX} and RES_{MIN}).
d) The differences are calculated as the result from our model minus the result from the ARMA(1,1) model, i.e. negative values signal better performance than the reference model.

The different aspects covered by the implemented forecasting study reveal a good prediction quality of the pricing model in terms of the closed-end fund market prices in our sample. This motivates us to examine the second application on how the information behind the pricing model could be used to create abnormal return generating trading portfolios.

6.2 Implementing Trading Rules

6.2.1 Experimental Design

Several studies show that abnormal returns can be earned by following trading strategies based on the information content revealed in the empirical premia.[4] In the second application of the derived pricing model we follow the idea of exploiting the information content using the modeled dynamic premia. For the implemented trading rule study we split the available five year sample into a four year parameter estimation period and a one year out of sample implementation period similar to the setup of the forecasting study.

The filter strategies are based on the information content of the residual premium difference defined by $RES_{t,i} = \pi_{t,i} - \theta_i$ for a closed-end fund i on day t. For this residual value the theoretical mean-reversion property of the continuous premia indicate the following information:

$$RES_{t,i} \begin{cases} > \epsilon : & \text{buying indication,} \\ < -\epsilon : & \text{shortselling indication, and} \\ \text{else} : & \text{no indication.} \end{cases} \tag{6.1}$$

The selection criterion for going long or short in a closed-end fund is whether its value of the dynamic premium π_t lies above or below its long-run mean θ. Upon the realizations of the premium residual we then build equally weighted trading portfolios containing closed-end funds that are hold long, short or not at all. The implemented symmetric filter strategies are based on a minimum amount ϵ, and $-\epsilon$ respectively, of the residual $RES_{t,i}$ in order for closed-end fund i to be included in the trading portfolio. The trading portfolios are revised on a monthly basis where new closed-end funds are selected according to the selection criterion and already contained closed-end funds are excluded when their premium residual becomes zero.

[4] For the US market of closed-end funds see, for example, Thompson (1978), Anderson (1986), and Pontiff (1995).

In order to be able to examine how long the information implicit in the premium residuals lasts in the market we run the filter rules for two scenarios. We use the information content collected (i) on week t and (ii) on week $t-1$ to actually rebalance the trading portfolios on week t. Thereby, we allow for a time lag of up to one week between the decision to trade and the actual implementation of the trade.

To evaluate the outcomes of the trading portfolios we utilize the concept of abnormal returns.[5] Let $HP_{t,i}$ denote the log-return of closed-end fund i for a one week holding period. Define $AR_{t,i}$ as the abnormal return for security i on week t, i.e. the excess return over a benchmark return:

$$AR_{t,i} = HP_{t,i} - BM_t.$$

As risk adjusting benchmark BM_t for week t we use the cross-sectional average $BM_t = \frac{1}{N}\sum_{i=1}^{N} HP_{t,i}$, with $N = 33$ closed-end funds. This risk adjustment for general market movements within the same market segment ensures that a naive buy-and-hold strategy of holding all closed-end funds in the sample over the fifth year earns a risk adjusted return of zero.[6] The amounts invested in each selected closed-end fund are assumed to be equal such that the abnormal return from the trading portfolio is given by the equally weighted average abnormal return defined by

$$\overline{AR}_t = \frac{1}{n_t}\sum_{i=1}^{n_t} AR_{t,i}\mathbf{1}_{RES_{t,i}},$$

$$\text{with } \mathbf{1}_{RES_{t,i}} = \begin{cases} +1 \text{ if filter strategy indicates buying,} \\ -1 \text{ if filter strategy indicates shortselling, and} \\ \ \ 0 \text{ otherwise} \end{cases}$$

across the number n_t of selected funds contained in the trading portfolios on week t. The average abnormal returns are then cumulated over the fifth year and the reported t-tests on a given trading strategy are computed as in Campbell, Lo, and MacKinlay (1997).

Analogously to the described trading strategies based on the information content of the dynamic premia we implement the same trading rules using the empirical values for the premia as defined in equation (4.1). For this benchmark test we use the residual of the empirical premium $RES_{t,i} =$

[5] See, for example, Bjerring, Lakonishok, and Vermaelen (1983), Peavy (1990), and Campbell, Lo, and MacKinlay (1997).

[6] The naive buy-and-hold strategy is similar to what serves Thompson (1978) as control strategy ('All Funds' trading strategy).

$PREM_{t,i} - \overline{PREM}_i$, where $\overline{PREM}_i$ serves us as proxy for the long-run mean value of the empirical premia and is calculated as the sample mean over the first four years. We then apply the selection criterion of equation (6.1) with opposite signs, since empirical discounts correspond to positive dynamic premia. The final procedure to calculate cumulated abnormal returns is done as described above.

6.2.2 Test Results on Trading Strategies

The results on the implemented filter strategies are reported separately for the two implementation lags as well as for the different trading strategies, i.e. the long, short and combined strategies that use the full information on long and short positions. The results based on the information content of the dynamic premia and the results for the benchmark test are shown, respectively, in tables 6.2 and 6.3 and can be summarized as follows.

From table 6.2 we first see that nearly all trading portfolios significantly earn positive cumulated abnormal returns (CARs) over the fifth year. Using no filter restriction, i.e. including all 33 funds, the immediate implementation of a combined strategy yields a risk adjusted cumulated abnormal return of 21.8 percent which marginally falls to 20.0 percent for an implementation lag of one week. Second, from the average number of observations (Obs.) for a given portfolio we see that the long strategies contain two to six times more closed-end funds in their trading portfolios on average than do the short strategies. This indicates that the dynamic premia lie above their long-run equilibrium θ in most cases of the fifth sample year which corresponds to observable closed-end fund discounts of the empirical premia. Third, we notice in general that the short strategies earn higher cumulated abnormal returns than do the buying strategies. With an immediate implementation of a long strategy the cumulated abnormal return results start from 13.6% and then gradually fall to negative values for higher filter values. However, the cumulated abnormal returns of around 13% are remained for filter values up to 0.1 which signal that the optimal filter height for a trading strategy lies between zero and ten percent. The comparable short strategy selects a daily average of 5.7 funds and yields a cumulated abnormal return of 58.3% with no filter implemented. Here, the highest cumulated abnormal return with 65.2% is realized for an intermediate filter height of 0.125. Comparing the overall results for the immediate and the lagged implementation of the trading strategies show that the main results are preserved and the information content persists over one week. The realizable values for the cumulated abnormal returns, however, are

Table 6.2. Results on Portfolio Trading Strategies

	Long Strategies				Short Strategies				Combined Strategies			
Filter	Obs.	nT	CAR	t-stat	Obs.	nT	CAR	t-stat	Obs.	nT	CAR	t-stat
(i) No Implementation Lag												
0	27.3	90	0.136	3.26***	5.7	42	0.583	8.92***	33.0	132	0.218	4.73***
0.025	25.8	92	0.138	3.32***	4.9	34	0.615	9.17***	30.7	126	0.222	4.82***
0.05	24.1	94	0.133	3.17***	4.3	30	0.552	8.09***	28.4	124	0.201	4.35***
0.075	21.5	80	0.126	2.96***	3.9	28	0.612	8.99***	25.4	108	0.199	4.24***
0.1	19.5	82	0.121	2.78***	3.6	30	0.566	8.44***	23.1	112	0.193	4.06***
0.125	16.0	86	0.102	2.31**	3.2	22	0.652	9.79***	19.1	108	0.183	3.76***
0.15	11.6	74	0.064	1.48*	2.7	18	0.586	8.19***	14.3	92	0.150	3.08***
0.175	9.4	60	0.055	1.24	2.6	22	0.449	6.50***	11.9	82	0.145	2.88***
0.2	7.3	48	0.009	0.19	2.4	18	0.380	5.53***	9.7	66	0.122	2.25**
0.225	6.2	34	-0.052	-1.12	2.0	14	0.340	6.25***	8.3	48	0.073	1.51*
0.25	5.1	30	0.046	0.98	2.0	14	0.328	5.95***	7.1	44	0.145	2.96***
0.275	4.3	24	-0.011	-0.22	1.9	12	0.349	6.31***	6.2	36	0.160	3.24***
0.3	3.9	22	-0.023	-0.48	1.6	10	0.330	6.02***	5.5	32	0.120	2.42***
(ii) Implementation Lag of One Week												
0	27.3	90	0.124	2.91***	5.7	42	0.557	8.35***	33.0	132	0.200	4.24***
0.025	25.8	92	0.116	2.76***	4.9	34	0.563	8.19***	30.7	126	0.194	4.12***
0.05	24.1	94	0.110	2.58***	4.3	30	0.574	8.04***	28.4	124	0.183	3.83***
0.075	21.5	80	0.093	2.16**	3.9	28	0.620	8.71***	25.4	108	0.175	3.62***
0.1	19.5	82	0.093	2.10**	3.6	30	0.561	7.90***	23.1	112	0.170	3.45***
0.125	16.0	86	0.062	1.39*	3.2	22	0.608	8.60***	19.1	108	0.147	2.92***
0.15	11.6	74	0.017	0.37	2.7	18	0.589	7.76***	14.3	92	0.112	2.20**
0.175	9.4	60	-0.003	-0.06	2.6	22	0.557	7.58***	11.9	82	0.111	2.09**
0.2	7.3	48	-0.061	-1.23	2.4	18	0.476	6.51***	9.7	66	0.069	1.19
0.225	6.2	34	-0.070	-1.45*	2.0	14	0.318	5.63***	8.3	48	0.053	1.06
0.25	5.1	30	0.007	0.14	2.0	14	0.298	5.29***	7.1	44	0.116	2.29**
0.275	4.3	24	-0.014	-0.28	1.9	12	0.305	5.39***	6.2	36	0.129	2.56***
0.3	3.9	22	-0.020	-0.41	1.6	10	0.312	5.57***	5.5	32	0.123	2.45***

Notes:
For each type of strategy we first report the average number of CEFs (Obs.) contained in the trading portfolios and the number of trades (nT) needed to realize the trading strategy. Statistically significant cumulated abnormal returns (CARs) at the 1%-, 5%- and 10%-levels are denoted by ***, **, and *, respectively.

Table 6.3. Benchmark Results of Trading Rules Based on Empirical Premia

	Long Strategies				Short Strategies				Combined Strategies			
Filter	Obs.	nT	CAR	t-stat	Obs.	nT	CAR	t-stat	Obs.	nT	CAR	t-stat
(i) No Implementation Lag												
0	27.5	94	0.132	3.07***	5.5	44	0.607	9.70***	33.0	138	0.213	4.59***
0.025	26.5	82	0.140	3.35***	4.5	32	0.649	9.85***	31.0	114	0.214	4.69***
0.05	25.3	72	0.133	3.18***	3.5	20	0.731	10.24***	28.8	92	0.208	4.54***
0.075	23.0	58	0.129	3.07***	3.3	18	0.686	9.44***	26.3	76	0.205	4.43***
0.1	22.0	54	0.144	3.43***	3.0	14	0.623	8.16***	25.0	68	0.211	4.56***
0.125	20.5	48	0.136	3.48***	2.8	14	0.557	7.62***	23.3	62	0.193	4.41***
0.15	17.8	44	0.146	3.72***	2.7	12	0.536	7.70***	20.5	56	0.199	4.62***
0.175	14.7	44	0.128	3.17***	2.6	12	0.504	7.64***	17.3	56	0.183	4.14***
0.2	9.1	36	0.148	3.57***	2.2	8	0.490	7.61***	11.3	44	0.213	4.60***
0.225	5.5	28	0.078	1.73**	1.8	8	0.481	7.36***	7.3	36	0.167	3.30***
0.25	4.3	18	0.084	1.86**	1.8	8	0.481	7.36***	6.1	26	0.199	3.83***
0.275	2.5	12	-0.064	-1.29	1.8	8	0.481	7.36***	4.3	20	0.152	2.60***
0.3	1.3	4	-0.011	-0.22	1.1	4	0.421	6.71***	2.4	8	0.210	3.58***
(ii) Implementation Lag of One Week												
0	27.5	94	0.116	2.67***	5.5	44	0.580	8.75***	33.0	138	0.188	3.95***
0.025	26.5	82	0.130	3.06***	4.5	32	0.640	9.24***	31.0	114	0.199	4.23***
0.05	25.3	72	0.120	2.83***	3.5	20	0.616	8.39***	28.8	92	0.182	3.90***
0.075	23.0	58	0.122	2.88***	3.3	18	0.566	7.57***	26.3	76	0.184	3.90***
0.1	22.0	54	0.137	3.26***	3.0	14	0.533	6.69***	25.0	68	0.197	4.15***
0.125	20.5	48	0.135	3.43***	2.8	14	0.520	6.80***	23.3	62	0.189	4.20***
0.15	17.8	44	0.141	3.57***	2.7	12	0.539	7.48***	20.5	56	0.200	4.56***
0.175	14.7	44	0.124	3.07***	2.6	12	0.533	7.75***	17.3	56	0.189	4.17***
0.2	9.1	36	0.144	3.45***	2.2	8	0.575	8.61***	11.3	44	0.231	4.89***
0.225	5.5	28	0.044	0.96	1.8	8	0.560	8.15***	7.3	36	0.163	3.13***
0.25	4.3	18	0.022	0.48	1.8	8	0.560	8.15***	6.1	26	0.177	3.35***
0.275	2.5	12	-0.075	-1.53*	1.8	8	0.560	8.15***	4.3	20	0.176	2.94***
0.3	1.3	4	-0.082	-1.63*	1.1	4	0.505	7.67***	2.4	8	0.233	3.84***

Notes:
For each type of strategy we first report the average number of CEFs (Obs.) contained in the trading portfolios and the number of trades (nT) needed to realize the trading strategy. Statistically significant cumulated abnormal returns (CARs) at the 1%-, 5%- and 10%-levels are denoted by ***, **, and *, respectively.

not as high any more. Fourth, in implementing the portfolio strategies with the aim of earning high cumulated abnormal returns one needs to consider the transactions costs associated with these strategies. The reference number of trades (nT) is 66 which stems from building a buy-and-hold portfolio containing all closed-end funds and its liquidation at the end of the year.[7] In order to administer the long and the combined strategies the number of trades lies above 66 for attractive cumulated abnormal returns. The highly profitable short strategies can be followed with twenty to forty transactions. Compared to the short strategies we further find the long portfolio holdings more stable over time in that the relation of the number of trades to the included closed-end funds is generally lower. With the objective to lower transactions costs it seems optimal to pursue an intermediate filter strategy of around 0.1 which only marginally sacrifices the returns. Finally, in order to evaluate the outcomes of the trading test we compare the model based results to those obtained from the information content of the empirical premia which we report in table 6.3. The facts already described for the model based results principally hold for the benchmark results as well. When comparing the realizable cumulated abnormal returns our model performs better for filter values close to zero in terms of long and combined strategies. Overall, we see the similar pattern that the cumulated abnormal returns remain stable up to certain filter values while the number of closed-end funds and the number of transactions decrease.

[7] The return of this naive buy-and-hold strategy is already incorporated as benchmark return in our reported cumulated abnormal returns.

7

Summary and Conclusions

Part II of this book contains a theoretical and empirical analysis on the special equity asset class of closed-end funds. The analysis contributes to the pricing literature on closed-end funds in that we develop a valuation model that captures the distinct pricing characteristics of closed-end fund market shares. Given the pricing model we are able to perform a general closed-end fund analysis and derive insights into two potential applications of the model valuable for investment decisions.

In chapter 4 we focus on the theoretical modeling perspective where we derive a closed-form, two-factor valuation model for closed-end fund market shares which captures the distinct pricing characteristics of closed-end funds trading at discounts which vary over time. As explaining factors we utilize the underlying price of the NAV and a stochastic premium which is modeled with dynamic mean-reversion. Thereby, the pricing model incorporates both the price risk stemming from the originating portfolio investments and the risk associated with time varying values for the premia.

For the empirical implementation of the theoretical pricing model in chapter 5 we use a state space representation to estimate the model parameters and the time-series of the dynamic premium by maximum likelihood inferences. In the empirical estimations we use a historical sample of emerging market closed-end funds for the period of 1993 to 1997. For the examined closed-end fund sample we observe an average of the empirical discount of 3.25 percent which is consistent with the literature. The sample further confirms a higher volatility of the share prices than the prices of the underlying net asset value which is in line with the noise trader argument. In the empirical analysis of our pricing model we find a high mean-reversion of the dynamic premia with an approximate half-life of five months. This indicates high fluctuations of the premia over time which is emphasized by a premium volatility of 30 percent on average and the outcome we report for the time-series of the dynamic premia. For the results on the parameter estimates of the factor volatilities we obtain a higher volatility for the dynamic premia than for the net asset value on average. This presents evidence that the risk with altering premia is higher than the underlying price risk.

The possible use of the pricing model to investors is evaluated by inferring insights into two suggested applications in chapter 6. First, we implement an out of sample forecasting study in order to evaluate the prediction quality of the model in terms of closed-end fund market prices. The estimation results based on our sample document that the models performance is better in most aspects when compared to a well specified econometric reference model. In a second application we test whether the information contained in the dynamic premia can be used to build portfolio strategies with the objective to generate risk adjusted abnormal returns. In an out of sample setting we find that long trading strategies can earn cumulated abnormal returns of around fourteen percent which can be increased to over twenty percent following both long and short strategies. Further, we see that an immediate implementation of the strategies is more valuable than implementing them with a time lag which indicates that the information content of the premia vanishes over time. Overall, we find that the proposed valuation model to price closed-end funds can generate valuable information for investment purposes.

Part III

Pricing Fixed-Income Securites

8

Introduction and Survey

8.1 Overview

In part III we explore a new approach in modeling the term structure of interest rates and the pricing of fixed-income securities. The pricing model we present is a type within the affine class and is estimated using interest rate panel data. This first chapter clears the basic definitional terminology when dealing with bond prices and interest rates. We further clear our view on the enormous spectrum of evolved asset pricing models for fixed-income securities by performing a survey. In chapter 9 we motivate our model choice, present our dynamic term structure model, and derive a closed-form solution for the price of discount bonds as the security of primary interest. In the next chapter we perform a comparative statistic in order to evaluate the characteristic theoretical properties of our interest rate model. Here, we especially examine the distinct behavior of the term structure of interest rates as well as the term structure of volatilities which are relevant in valuing derivatives. The following chapter 11 is dedicated to the management of interest rate risk. Starting with a clarification of the types of risk involved, we further implement the very useful technique of a duration measure for our model, and finally show how to price term structure derivatives. Among the most relevant fixed-income instruments in risk management we specialize in valuing bond options, swap contracts, and interest rate floor and cap agreements. Chapter 12 considers the calibration of the proposed term structure model to standard fixed-income instruments. First, we describe the different econometric methods used in the literature to then specialize our implemented version that resembles the theoretical model properties at best. For the different interest rate markets we choose data from US government issues, LIBOR rates and swap rates. Upon these data we report our inferences on the model parameters and analyze the filtered state variables. Finally, we conclude in the last chapter with a summary of our findings.

Before we describe our proposed term structure model in the next chapter, we first take a look at the incomplete market for fixed-income securities. There, we unfortunately realize that interest rates are not directly tradeable. Instead different financial instruments are traded on the markets which depend on interest rates themselves. Guided by this fact, we initially consider the basic financial instruments in the bond market and how they are related to each other. Second, we closer examine the peculiarity of the risk premium concept that generally need to be taken into account in interest rate modeling. Last but not least, we theoretically motivate our choice of a special interest rate model with a risk premium following stochastic dynamics.

8.2 Bond Prices and Interest Rates

We start with the framework of calculating interest rates in continuous time. The framework of continuous compounding can be traced back to two different origins: (i) The field of *econometrics*[1], where in time-series analysis the differences in logarithmic security prices P_t, which can be equity prices, bond prices or foreign exchange rates,

$$r_t = \ln(P_t) - \ln(P_{t-1})$$

are called logarithmic yields r, or just "first differences" in terms of levels. The use in the field of finance lays (ii) in the *theory of finance* as described, for example in Ingersoll (1987), where the continuous compounding evolves from within period compounding. Instead of one compounding period, we incorporate n subperiods, and obtain for the limiting case of getting interest payments continuously

$$\frac{P_t}{P_{t-1}} = \lim_{n\to\infty}\left(1+\frac{r}{n}\right)^n;$$

this result can be stated in the form of the Taylor series expansion

$$\lim_{n\to\infty}\left(1+\frac{r}{n}\right)^n = 1+\frac{1}{1!}r+\frac{1}{2!}r^2+\frac{1}{3!}r^3+o\left(r^3\right) = e^r.$$

The equivalence between the use in econometrics and finance is given by the law of logarithms that $\ln(P_t) - \ln(P_{t-1}) = \ln(P_t/P_{t-1})$. More formally,

[1] See, for example, Campbell, Lo, and MacKinlay (1997).

we now define the basic building blocks of the bond market, consisting of the prices of pure discount bonds, coupon bonds, the money market account, and the interest rates, as follows.

Definition 8.1 (Bond Market) *A discount bond is defined as a non-defaultable security, which guarantees the holder to receive \$1 at a prespecified maturity date T; its price at time t will be denoted by $P(t,T)$. The theoretical price of a coupon bond $P^c(t,T)$ will then be*

$$P^c(t,T) = \sum_{u \in (t;T]} c(u) \cdot P(t,u)$$

where $c(u)$ denotes the cash flows (coupon and notional) received from the coupon bond holder at time u.[2]

The relationship between discount and coupon bond prices is rather crucial, because in real world capital markets discount bonds are rarely available.[3] Instead, we are unluckily stuck with extracting our necessary information from market prices of traded coupon bonds, which even might be additionally equipped with special features that make the analysis even worse.[4] A common way to extract the term structure of interest rates is to use market prices of coupon bonds, interest rate futures, or interest rate swaps.[5] In the latter case we reinterpret a coupon bond in light of the portfolio theory as a portfolio of distinct cash flows discounted with their corresponding interest rates which we define next.

Definition 8.2 (Interest Rates) *Given the price $P(t,T)$ of a discount bond we define the continuously compounded spot rates for the period $[t,T]$ by*

---

[2] Otherwise there would exist arbitrage opportunities of Type 2, according to the nomenclature of Ingersoll (1987). Farkas's Lemma (see Farkas (1902)) can serve as criterion for this arbitrage-free relationship between discount and coupon bond prices.

[3] However, stripping coupon bonds into separate cash flows is becoming more popular and widespread in the US and Germany. For an empirical analysis of the US Treasury STRIPS program see, for example, Love and McShea (1997) and Grinblatt and Schwartz (2000).

[4] For example, a bond with embedded options such as callable US Treasury Bonds, will probably deteriorate from the price of an equivalent pure coupon bond (Bliss and Ronn (1998)).

[5] See, for example, Bliss (1997) and Uhrig and Walter (1996).

$$y(t,T) = -\frac{\ln P(t,T)}{T-t}. \tag{8.1}$$

The functional mapping $T \longmapsto y(t,T)$, for a given time t, is called the yield curve. The spot rate $y(t,T)$ collapses at the short end of the term structure to

$$r(t) = \lim_{T \downarrow t} -\frac{\ln P(t,T)}{T-t} = -\left.\frac{\partial P(t,t')}{\partial t'}\right|_{t'=t} = f(t,t),$$

which equivalently is the instantaneous short and forward rate. The general forward rate for $[T_1, T_2]$ contracted at t is defined as

$$f(t,T_1,T_2) = -\frac{\ln P(t,T_2) - \ln P(t,T_1)}{T_2 - T_1}. \tag{8.2}$$

By taking the limit of $T_2 \downarrow T_1$, we converge at the instantaneous forward rate for the maturity T contracted at t

$$f(t,T) = -\frac{\partial \ln P(t,T)}{\partial T},$$

which constitutes the rate for instantaneous borrowing and lending in T.

By these definitions on interest rates, we can characterize the relationship between the term structure of spot and forward rates. We differentiate the formula for the spot rates (8.1) with respect to T:

$$\frac{\partial R(t,T)}{\partial T} = -\left(\frac{\partial \ln P(t,T)}{\partial T}\frac{1}{T-t} - \ln P(t,T)\frac{1}{(T-t)^2}\right).$$

Hence, we are able to deduce the rearrangement

$$f(t,T) = y(t,T) + (T-t)\frac{\partial R(t,T)}{\partial T}$$

which shows that the forward curve superficially resembles the spot curve dependent on the slope of the spot curve. The forward curve is higher than the spot curve if the yield curve is increasing, and vice versa. Moreover, using the definitions on interest rates we gain the ordinary but crucial relationships

$$\begin{aligned} P(t,u) &= e^{-(u-t)\cdot r(t,u)}, \text{ and} \\ P^c(t,T) &= \sum_{u \in (t;T]} CF(u) \cdot e^{-(u-t)\cdot r(t,u)} \end{aligned}$$

for discount and coupon bond prices, respectively. By restating the price of a discount bond as

$$P(t,T) = \exp \int_t^T f(t,\zeta)\, d\zeta$$

we see that we need to take information on future interest rates or the forward rate into account for a fair valuation of bonds. Equivalently, market prices on bonds reveal information about these interest rates.

A different security which will be of interest as so-called numéraire asset is the money market account as defined in section 1.2.1. This security looks into the past and present information on interest rates. Finally, to value financial instruments such as interest rate swaps, cap or floor agreements we introduce a special interest rate, the LIBOR rate. This interest rate uses simple instead of continuous compounding as we see from its definition.

Definition 8.3 (LIBOR Rate) *The simple spot rate or LIBOR rate is defined as*

$$L(t,T) = -\frac{P(t,T) - 1}{(T-t)\, P(t,T)} \tag{8.3}$$

for the time interval $[t,T]$.

The relationship between interest rates and forward rates shows that interest rates of different maturities are heavily correlated.[6] Therefore and to keep an interest rate model arbitrage-free and analytically tractable, the stochastic processes governing $r(t)$ or $f(t,T)$ are assumed to be functions of a fixed number of Brownian motions.

8.3 Dynamic Term Structure Models

In order to help integrate our proposed model from the next chapter in the spectrum of current asset pricing models for fixed-income securities we perform a survey in this section. The era of dynamic term structure models starts with the re-examination of the traditional hypotheses of the term structure of interest rates - such as the *Expectations Hypothesis*, the *Liquidity Theory* of Hicks (1939) , and the *Preferred Habitat Theory* of Modigliani and Sutch (1966) - by Cox, Ingersoll, and Ross (1981a) in a dynamic continuous-time context. Since the early models marked by the seminal works of Vasicek (1977) and Cox, Ingersoll, and Ross (1985b) there evolved an enormous body

[6] If one would implement a multi factor regression, this would possibly give rise to a multicolinnearity problem. See the analysis of chapter 12.

of dynamic asset pricing models designed for pricing fixed-income securities. For an in-depth treatment and the latest advances of modeling and estimation issues see, for example, Musiela and Rutkowski (1997), Chapman and Pearson (2001), Brigo and Mercurio (2001), Dai and Singleton (2003a), and Dai and Singleton (2003b). In the following survey we choose to categorize term structure models into three different classes - the *affine model class*, the *no-arbitrage models*, and the *random field models* - which covers the most important developments in term structure modeling. Thereby, we concentrate on dynamic term structure models of *default-free bond yields*. Thus for issues on credit risk modeling see, for example, the detailed overviews of Bielecki and Rutkowski (2001) and Schoenbucher (2003) and references therein. For recent research analysis on both 'reduced-form' and 'structural' models and results on determinants of and evidence on credit spreads see, for example, Collin-Dufresne and Goldstein (2001), Eom, Helwege, and Huang (2002), Huang and Huang (2003), and Dai and Singleton (2003b).

The *affine class of models* of Duffie and Kan (1996) has become the dominant framework for modeling the term structure of interest rates.[7] This is probably partly due to their eminent analytical tractability in terms of pricing various fixed-income securities, the richness of realizable yield distributions, and the ease of empirical implementation with several available estimation methods. The idea of affine models is to specify a number of (un)observable state variables or risk factors and map these on the short-term interest rate and the factor risk premia, which can conveniently be done in an exponentially affine framework. For an detailed overview of affine models in terms of modeling aspects, estimation methods, empirical evidence, and joint systems see, for example, Dai and Singleton (2003b) and Piazzesi (2003). Since our proposed model belongs to this affine class, we provide further insight into modeling and estimation aspects of this class, respectively, in chapters 9 and 12.

The class of *no-arbitrage models* has been pioneered by the works of Ho and Lee (1986) and Heath, Jarrow, and Morton (1992). In contrast to the more time-series oriented affine models, the no-arbitrage models take the entire observable yield curve as given to ensure consistency with the market. Thus these models have a cross-sectional focus which is often mandatory in the financial industry for practical demands of point-in-time pricing systems. From there, arbitrage-free movements of the interest rates or bond prices over

[7] For further developments of the affine class see, for example, Dai and Singleton (2000) who introduce a canonical model, and Duffee (2002) and Duarte (2003) who propose more flexible risk preference specifications.

time are modeled. However, the models typically implemented in the industry are silent about the inherent time-series behavior of yields under the physical measure. An important recent development in the class of no-arbitrage models are the so-called *market models*, based on the work by Brace, Gatarek, and Musiela (1997), Miltersen, Sandmann, and Sondermann (1997), and Jamshidian (1997). This market approach constructs arbitrage-free models for forward LIBOR and swap rates at an observed discrete tenor. The pricing of fixed-income securities such as caps, floors, and swaptions is based on forward measures, with the numéraire being a LIBOR based discount bond or a swap price. If two derivatives based on the same underlying risk factors are priced with different numéraires, then the question arises if the resulting pricing models may be based on mutually inconsistent assumptions about the risk factor distributions. Finally, we note that the practice of point-in-time calibration of no-arbitrage models is not without controversy. Recalibrating the model parameters when the shape of the underlying yield curve or option volatilities change amounts to 'changing the model', i.e. from a dynamic perspective they are almost surely fraught with arbitrage opportunities. An extended treatment of the no-arbitrage models is found in Musiela and Rutkowski (1997) and current issues in, for example, Dai and Singleton (2003a).

The *random field models* are build upon the original insights of Heath, Jarrow, and Morton (1992) yielding high and possible indefinite dimensional models. Such high dimensional model, developed under the labels of 'Brownian Sheets' (Kennedy (1994) and Kennedy (1997)), 'Random Fields' (Goldstein (2000)) and 'Stochastic String Shocks' (Santa-Clara and Sornette (2001)). These models view the entire yield curve as the state vector and explicitly model its dynamic properties, thereby, in principle, the problematic issues of the no-arbitrage models mentioned above disappear. A further development of these models is given by Kimmel (2003) who introduces random field models containing stochastic volatility. For aspects of empirically estimating random field models, Collin-Dufresne and Goldstein (2003) point out that random fields being infinite-factor models, i.e. they cannot possess a finite state space representation, are considered mostly not implementable unless the framework is approximated as a low dimensional system. A tractable framework is given by specifying discrete forward bond price dynamics which is empirically investigated by the papers of Longstaff, Santa-Clara, and Schwartz (2001a), Longstaff, Santa-Clara, and Schwartz (2001b), and Han (2003). There the authors initially specify a random field model, but effectively implement a low dimensional factor approximation; Kerkhof and Pelsser (2002) emphasizes that such models are observationally equivalent to finite factor market models.

The paper by Collin-Dufresne and Goldstein (2003) makes an effort of *unifying the model classes.* They identify a generalized class of term structure models which significantly extends the affine class studied by Duffie, Pan, and Singleton (2000) and Chacko and Das (2002). This class of models can accommodate both *infinite-state variable models* (i.e. Heath, Jarrow, and Morton (1992) type of models) as well as *infinite-factor models* (i.e. the random field models) and possesses analytic solutions for the characteristic function. This enables them to provide closed-form solutions for many types of fixed-income securities. Further promising results of this pricing framework is, for example, reported in the empirical analysis of Han (2003). He captures the relative pricing of caps and swaptions developing and estimating a model with four yield factors and three covariance factors. For an empirical comparison of affine models and random field models see, for example, Bester (2003) who develops a state space framework using Markov-Chain-Monte-Carlo algorithms and a panel of forward rate data.

9

Term Structure Model

9.1 Modeling an Incomplete Market

There exists some general aspects of using a factor model within the affine class of Duffie and Kan (1996) to price interest rate derivatives. In modeling bond markets in section 1.2.1 we examined the following interrelated peculiarities: the incompleteness of the market, the maturity independent risk premium, and the pricing of interest rate derivatives. To study these aspects we used a single-factor model that chooses the instantaneous short rate as the state variable:

> "More recently it has been recognized that, if assumptions are made about the stochastic evolution of the instantaneous rate of interest in a continuous time model, much richer theories of bond pricing can be derived, which constrain the relationship between the risk premia on bonds of different maturities."[1]

Intuitively, the choice of the short rate is a natural starting point in that the price of a discount bond $P(t,T)$ depends on the behavior of the short rate during the interval $[t,T]$ as we demonstrated in the previous section. From an empirical point of view, statistical approaches to interest rate modeling adopted and followed by, for example, Litterman, Scheinkman, and Weiss (1991) and Rebonato (1998) show that the first factor can be attributed with the level of interest rates. In fact, this is the classical approach to interest rate theory adopted in the field of financial economics with the most prominent representatives being the arbitrage-free Vasicek (1977) model and the utility based methodology of Cox, Ingersoll, and Ross (1985b). The models make assumptions about the stochastic evolution of the short rate in a continuous

[1] Brennan and Schwartz (1979, p. 134).

time model. The specification of such dynamics are chosen with the aim of incorporating the fluctuations of the short rate during time at best. The peculiarity about these types of models is that the short rate as chosen factor is a non-traded underlying object in the interest rate market. Therefore, the bond market is an incomplete market. In section 1.2.1 we saw that we need to employ a market price of risk parameter in order to price fixed-income securities. At this point we see the possibility to establish a new dimension in term structure modeling in that we build a term structure model that captures not a time homogeneous but a time varying market price of risk $\lambda(t)$.

9.2 Motivation for a Stochastic Risk Premium

In financial economics we model investors' attitudes towards risk in order to come up with a reasonable relationship of risk and return in the relevant market. In standard microeconomic theory investors' *tastes* are modeled by specifying their utility function, being, for example, an investor with a logarithmic utility function as in the production-exchange economy of Cox, Ingersoll, and Ross (1985a). To get a utility function for a representative investor, we then aggregate the utility functions across investors with potential aggregation problems.[2] In an equilibrium term structure model this utility function indirectly enters into the form of the market price of risk. However, looking at real world security markets we see different investors present on different trading days. Thus, we could model these changing representative tastes over time by allowing for a dynamic behavior of the market price of risk $\lambda(t)$. This approach is inspired by a shift from modeling static tastes as with previous term structure models to capture the *beliefs* of the dynamic behavior of tastes. Instead of assuming the same representative investor being present over time, i.e. assuming $\lambda(t) = const$, we set up a plausible dynamic model of how the market price of risk possibly fluctuates over time.

Besides the purpose of modeling the dynamic behavior of investors' tastes we additionally want to build a term structure model that has promising features in comparison to extant models. A range of desirable theoretical, technical, and empirical features can be summarized by requiring some criteria that need to be met by the model. For Rogers (1995) a term structure model should be:

"(*a*) *flexible* enough to cover most situations in practice; (*b*) *simple* enough that one can compute answers in reasonable time; (*c*) *well-*

[2] See Ingersoll (1987).

specified, in that required inputs can be observed or estimated; (*d*) *realistic*, in that the model will not do silly things. Additionally, the practitioner shares the view of an econometrician who wants (*e*) *a good fit* of the model to data; and a theoretical economist would also require (*f*) *an equilibrium derivation* of the model."[3]

A class of term structure models that broadly meet these criteria and has recently been attended considerable focus is the *affine class*. This class is defined by factor models in which the drift and volatility coefficients of the state variable processes are affine functions of the underlying state vector.[4] Our approach is to work within this class of factor models and extend the single-factor term structure models, such as the model used in the previous section, to include the market price of risk $\lambda(t)$ as a second state variable. For the relevance and a detailed study of such a broader class of affine models which allow for time-varying risk premia see, for example, Dai and Singleton (2002), Duffee (2002), and Duarte (2003). We introduce a *two-factor model* that enables us to analyze the investors' beliefs with respect to the interest rate risk inherent in the observable term structures on the markets. We aim at building a two-factor model that accommodates the above mentioned criteria. Especially, as we will see in the following section, we try to build a maximal model that does not suffer from implicitly imposing potentially strong over-identifying restrictions.[5]

Looking at the extant financial literature on this subject we already find several propositions of two-factor model specifications. The main idea is that single-factor models based on specifying a dynamic structure of the short interest rate exhibit some shortcomings that are rather unrealistic.[6] The most problematic feature of such models is that they endogenously result in a perfect correlation among bond prices of different maturities. Several studies have shown that such single-factor models are not suitable in modeling the term structure of interest rates.[7]

An overview of proposed two-factor specifications of term structure models is given in table 9.1. The first extensions were to incorporate the shape

[3] Rogers (1995, p. 93).

[4] See, for example, Duffie and Kan (1996).

[5] See, for example, the specification analysis of Dai and Singleton (2000).

[6] See, for example, Chen (1996b, ch. 2).

[7] The empirical studies on this subject include Chen and Scott (1992) and Pearson and Sun (1994) who use a maximum likelihood estimation, the works by Heston (1992) and Gibbons and Ramaswamy (1993) based on the generalized methods of moments, and the factor analysis of Litterman and Scheinkman (1991).

Table 9.1. Developments in Two-Factor Models

Authors	Model Specification	Constants
Brennan/ Schwartz (1979,80,82)	$dr = (a_1 + b_1 (l - r))\,dt + r\sigma_1 dz_1$ $dl = l\,(a_2 + b_2 r + c_2 l)\,dt + l\sigma_2 dz_2$ $dz_1 dz_2 = \rho dt$	$a_1, b_1, \sigma_1,$ $a_2, b_2, c_2, \sigma_2,$ ρ
Schaefer/ Schwartz (1984)	$ds = m\,(\mu - s)\,dt + \gamma dz_1$ $dl = \left(\sigma^2 - ls\right) dt + \sigma\sqrt{l} dz_2$ $dz_1 dz_2 = \rho dt$	$m, \mu, \gamma,$ $\sigma,$ ρ
Pennacchi (1991)	$dr = (a_1 + b_{11}(r + \sigma_k) + b_{12}\pi)\,dt + \sigma_r dz_r$ $d\pi = (a_2 + b_{21}(r + \sigma_k) + b_{22}\pi)\,dt + \sigma_\pi dz_\pi$ $dz_r dz_\pi = \rho dt$	$a_1, b_{11}, b_{12}, \sigma_k, \sigma_r$ $a_2, b_{21}, b_{22}, \sigma_\pi,$ ρ
Beaglehole/ Tenney (1991)	$dr = \kappa_1 (\theta - r)\,dt + \sigma_1 dw_{1,t}$ $d\theta = \kappa_2 (\theta_2 - \theta)\,dt + \sigma_2 dw_{2,t}$ $dw_{1,t} dw_{2,t} = \rho dt$	$\kappa_1, \sigma_1,$ $\kappa_2, \theta_2, \sigma_2,$ ρ
Longstaff/ Schwartz (1992)	$dX = (a - bX)\,dt + c\sqrt{X} dZ_1$ $dY = (d - eY)\,dt + f\sqrt{Y} dZ_2$ $dZ_1 dZ_2 = 0$	$a, b, c,$ d, e, f

of the yield curve in some way. The ideas of Brennan and Schwartz (1979), Brennan and Schwartz (1980), Brennan and Schwartz (1982), and Schaefer and Schwartz (1984) were to model the processes for the long-term interest rate l in addition to the short rate of, respectively, r and s. In the context of an equilibrium asset pricing model Pennacchi (1991) uses the instantaneous real interest rate r along with the expected inflation π as factors to establish a nominal term structure model. The term structure model of Beaglehole and Tenney (1991) corresponds to an economy where the short rate r reverts to a drift rate θ which itself reverts to a fixed mean rate over time. From a specification point of view we will see in the next section that our model has similarities with those derived by Pennacchi (1991) and Beaglehole and Tenney (1991). Another approach is the idea to model the stochastic volatility of the short rate as a factor. In this line Longstaff and Schwartz (1992) extend the general equilibrium term structure model of Cox, Ingersoll, and Ross (1985b). Based on two orthogonal economic factors X and Y they are able to derive a term structure model which utilizes the short rate and the volatility of the short rate as factors. The idea to further add an additional third factor to the term structure models has also been followed in the literature.[8] However, the specification analysis performed by Dai and Singleton (2000) raises doubts on the enhanced richness of such models.

9.3 Economic Model

The proposed model belongs to the class of dynamic factor models. This model class has two distinguished implications for valuation purposes that can be inferred from the description of the theoretical valuation of these models: the *time-series implications* from the specification of the dynamic evolution of the chosen factors during time, and the *cross-section dimension* with a model for the functional relationship of the value of discount bonds and time to maturity at a single point of time. Thus, the parameters of the models fulfill two tasks. First, they determine the form of the random evolution of the factors through time by the specific form of the stochastic processes. Second, they fix the various types of term structures out of the realizable spectrum of different time to maturities. In the following we focus on the first aspect whereas the consecutive two sections focus on the second aspect and will clarify the distinction between the two dimensions.

To derive the term structure model we start with the stochastic dynamics of the chosen underlying factors from a time-series perspective. The idea is to extend the specification of the short rate $r(t)$ as an Ornstein-Uhlenbeck process[9] by a stochastic market price of risk $\lambda(t)$.

Assumption 9.1 (State Variables) *Under the objective probability measure* $\mathbb{P}$ *(where increments of Brownian motions are denoted as* $dW^{\mathbb{P}}$*) we specify the stochastic dynamics for the state variables* $\mathbf{x} = [r(t), \lambda(t)]'$*, with* $r(t)$ *being the instantaneous interest rate and* $\lambda(t)$ *the market price of the interest rate risk, as:*

$$d\mathbf{x} = \left[\boldsymbol{\mu}^{\mathbb{P}} + \mathbf{T}^{\mathbb{P}}\mathbf{x}\right] dt + \boldsymbol{\sigma} d\mathbf{W}_t^{\mathbb{P}}, \tag{9.1}$$

$$\text{with } \boldsymbol{\mu}^{\mathbb{P}} = [\mu_r, \mu_\lambda]', \qquad \mathbf{T}^{\mathbb{P}} = \begin{bmatrix} \kappa_r & 0 \\ \kappa_\lambda & \theta_\lambda \end{bmatrix},$$

$$\boldsymbol{\sigma} = \begin{bmatrix} \sigma_r & 0 \\ 0 & \sigma_\lambda \end{bmatrix}, \text{ and } d\mathbf{W}_t^{\mathbb{P}} = \left[dW_{r,t}^{\mathbb{P}}, dW_{\lambda,t}^{\mathbb{P}}\right]'.$$

---

[8] Chen (1996a), for example, develops a model using the short rate, the short-term mean, and the short-term volatility as explaining factors.

[9] Proposed by Vasicek (1977). For the reasonability of a mean-reversion specification for the short rate see, for example, Backus, Foresi, and Zin (1998).

Therein, we allow for a correlation of the two sources of uncertainty $dW^{\mathbb{P}}_{r,t}dW^{\mathbb{P}}_{\lambda,t} = \rho dt$ *between the short rate process* $r(t)$ *and the price process for* $\lambda(t)$. *This process specification assumes a mean-reversion for the short rate as well as for the market price of risk.*

Corollary 9.2 (Equivalent Martingale Measure) *Going through the Girsanov transformations* $W^{\mathbb{P}}_{r,t} = W^{\mathbb{Q}}_{r,t} + \int_0^t \lambda(\zeta)\,d\zeta$ *with the Girsanov Kernel* $\lambda(t)$, *which also holds for our stochastic specification of the drift* $\lambda(t)$, *and* $W^{\mathbb{P}}_{\lambda,t} = W^{\mathbb{Q}}_{\lambda,t} + \int_0^t \lambda_2 d\zeta$, *we obtain under the equivalent martingale measure* $\mathbb{Q}$ *(with Brownian motions* $W^{\mathbb{Q}}$*) for equations (9.1) the risk adjusted processes:*

$$d\mathbf{x} = \left[\boldsymbol{\mu}^{\mathbb{Q}} + \mathbf{T}^{\mathbb{Q}}\mathbf{x}\right]dt + \boldsymbol{\sigma} d\mathbf{W}^{\mathbb{Q}}_t, \tag{9.2}$$

$$\text{with } \boldsymbol{\mu}^{\mathbb{Q}} = \left[\mu_r, \mu_\lambda - \sigma_\lambda\lambda_2\right]', \qquad \mathbf{T}^{\mathbb{Q}} = \begin{bmatrix} \kappa_r & -\sigma_r \\ \kappa_\lambda & \theta_\lambda \end{bmatrix},$$

$$\boldsymbol{\sigma} = \begin{bmatrix} \sigma_r & 0 \\ 0 & \sigma_\lambda \end{bmatrix}, \qquad \text{and } d\mathbf{W}^{\mathbb{Q}}_t = \left[dW^{\mathbb{Q}}_{r,t}, dW^{\mathbb{Q}}_{\lambda,t}\right]'.$$

Under the drift transformations the mean-reversion property and the correlation $dW^{\mathbb{Q}}_{r,t}dW^{\mathbb{Q}}_{\lambda,t} = \rho dt$ *of the two processes are retained.*

Remark 9.3 (Full System Matrix) *Note that the system of stochastic differential equations (9.2) incorporates a full system matrix* $\mathbf{T}^{\mathbb{Q}}$, *i.e. we model a two-fold interdependence: The two modeled state variables* $r(t)$ *and* $\lambda(t)$ *may both depend on their own value as well as on the other factor's value. Furthermore, the state variables are related through the correlation between the Brownian increments.*

First, we apply Itô's formula on a general, not yet specified interest rate derivative security with price $P(t,T,\mathbf{x})$ on date t and a maturity date T based on the risk adjusted stochastic differential equations of equation (9.2).[10] This results in the following stochastic differential:

[10] The derivative security price $P(t,T,\mathbf{x})$ is a function of the variables time t, the short rate r, and the market price of risk λ, whereas the maturity date T is regarded as a parameter.

$$\frac{dP(t,T,\mathbf{x})}{P(t,T,\mathbf{x})} = \mu_P dt + \boldsymbol{\eta} d\mathbf{W}_t^{\mathbb{Q}}, \tag{9.3}$$

$$\begin{aligned}
\text{with } \mu_P &= \frac{1}{P(t,T,\mathbf{x})}\left[\frac{1}{2}\sigma_r^2 \frac{\partial^2 P(t,T,\mathbf{x})}{\partial r^2} + \rho\sigma_r\sigma_\lambda \frac{\partial^2 P(t,T,\mathbf{x})}{\partial r \partial \lambda}\right.\\
&\quad + \frac{1}{2}\sigma_\lambda^2 \frac{\partial^2 P(t,T,\mathbf{x})}{\partial \lambda^2} + \mu_1 \frac{\partial P(t,T,\mathbf{x})}{\partial r}\\
&\quad \left. + \mu_2 \frac{\partial P(t,T,\mathbf{x})}{\partial \lambda} + \frac{\partial P(t,T,\mathbf{x})}{\partial t}\right],\\
\mu_1 &= \mu_r + \kappa_r r - \sigma_r \lambda,\\
\mu_2 &= \mu_\lambda - \sigma_\lambda \lambda_2 + \kappa_\lambda r - \theta_\lambda \lambda, \text{ and}\\
\boldsymbol{\eta} &= \frac{1}{P(t,T,\mathbf{x})}\left[\sigma_r \frac{\partial P(t,T,\mathbf{x})}{\partial r} \quad \sigma_\lambda \frac{\partial P(t,T,\mathbf{x})}{\partial \lambda}\right]'.
\end{aligned}$$

To get to the fundamental partial differential equation, we use the *standard finance condition*[11] as a necessary condition[12] to ensure the pricing of the derivative in absence of arbitrage opportunities: In a risk-adjusted environment all financial claims have to earn the same rate of return as a riskless asset, i.e. the drift of the price process μ_P needs to equal the riskless interest rate $r(t)$. This approach is equal to Merton's argument that a riskless security or portfolio must earn the spot rate $r(t)$ in equilibrium.[13] Thus, we have a partial equilibrium condition which implies that there are no possibilities for arbitrage gains.[14] This leads us in our case to the following partial differential equation:

$$\begin{aligned}
&\tfrac{1}{2}\sigma_r^2 \tfrac{\partial^2 P(t,T,\mathbf{x})}{\partial r^2} + \rho\sigma_r\sigma_\lambda \tfrac{\partial^2 P(t,T,\mathbf{x})}{\partial r \partial \lambda} + \tfrac{1}{2}\sigma_\lambda^2 \tfrac{\partial^2 P(t,T,\mathbf{x})}{\partial \lambda^2}\\
&+\mu_1 \tfrac{\partial P(t,T,\mathbf{x})}{\partial r} + \mu_2 \tfrac{\partial P(t,T,\mathbf{x})}{\partial \lambda} + \tfrac{\partial P(t,T,\mathbf{x})}{\partial t} - rP(t,T,\mathbf{x}) = 0,
\end{aligned} \tag{9.4}$$

ordered in terms of derivatives. Equation (9.4) is called the *fundamental partial differential equation* in asset pricing since it holds for any fixed-income instrument P. Based on this equation we derive arbitrage-free prices of discount bonds in the following chapter in order to evaluate the theoretical properties of the implied term structures. With the obtained cross-section we then fully

---

[11] See, for example, Heath, Jarrow, and Morton (1992, pp. 86-87).

[12] See Morton (1988).

[13] I.e. the stochastic part in the models corresponding stochastic differential equation drops out. See, for example, Merton (1973).

[14] Note that the vice versa argument does not hold; freedom of arbitrage does *not* imply an equilibrium.

specify the dimensions of the term structure model besides the already given time-series properties. In the chapter thereafter we use equation (9.4) to price several fixed-income instruments and derivative securities that are crucial in risk management.

10

Initial Characteristic Results

In this chapter we present initial characteristic results of our term structure model. At first, we show how to price discount bonds considered to be the most basic interest rate contracts serving as pure securities in the sense of Arrow/Debreu. Based on the discount bond pricing formula we are able to derive the term structures of interest rates and volatilities. Further, we analyze limiting cases of our model where we first demonstrate that our model contains the Ornstein-Uhlenbeck process model proposed by Vasicek (1977) as special case. Second, we examine the asymptotic behavior of the term structure of spot interest rates at the short and long end. In the last section we discuss the core influences of the state variables and the model parameters on the shape of the term structures in a comparative statistic analysis. Such an analysis is considered relevant in order to know which type of term structures are realizable within the model and to get an idea on how changes in the values of the state variables and the model parameters influence the shape of the term structures.

10.1 Valuing Discount Bonds

In order to value discount bonds we specialize the partial differential equation (9.4) on the case of P being the value of a discount bond. Thereby, we impose the boundary of $P(T,T,\mathbf{x}) = 1$ for the discount bond price at maturity.

Proposition 10.1 (Discount Bond Price) *For our Gaussian term structure model we propose in light of Duffie and Kan (1996) an exponential affine structure for the discount bond prices* $P(t,T,\mathbf{x})$ *as:*

$$P(t,T,\mathbf{x}) = e^{-[A(t),B(t)]\mathbf{x}-C(t)}. \tag{10.1}$$

For the boundary conditions we need to fulfill $A(T) = B(T) = C(T) = 0$ *in order to ensure a final discount bond price of one at maturity. With this functional mapping of the short rate and its market price of risk we only need to calculate the functions* $A(t)$, $B(t)$, *and* $C(t)$ *in order to fully specify the cross-section dimension of our model.*

To prove the proposed solution we need to show that the discount bond price from equation (10.1) indeed solves the partial differential equation (9.4). Utilizing the standard *separation of variables* technique, as used for example by Chen (1995), we first calculate the partial derivatives and denote them by subindexes as

$$P_r = \frac{\partial P(t,T,\mathbf{x})}{\partial r} = -AP, \quad P_{rr} = \frac{\partial^2 P(t,T,\mathbf{x})}{\partial r^2} = A^2 P,$$

$$P_\lambda = \frac{\partial P(t,T,\mathbf{x})}{\partial \lambda} = -BP, \quad P_{\lambda\lambda} = \frac{\partial^2 P(t,T,\mathbf{x})}{\partial \lambda^2} = B^2 P, \text{ and}$$

$$P_{r\lambda} = \frac{\partial^2 P(t,T,\mathbf{x})}{\partial r \partial \lambda} = ABP, \; P_t = \frac{\partial P(t,T,\mathbf{x})}{\partial t} = (-A_t r - B_t \lambda - C_t) P,$$

which we insert into equation (9.4) to yield

$$\tfrac{1}{2}\sigma_r^2 A^2 + \rho\sigma_r\sigma_\lambda AB + \tfrac{1}{2}\sigma_\lambda^2 B^2 - (\mu_r + \kappa_r r - \sigma_r \lambda) A$$

$$- (\mu_\lambda - \sigma_\lambda \lambda_2 + \kappa_\lambda r t + \theta_\lambda \lambda) B - A_t r - B_t \lambda - C_t - r = 0.$$

We then separate for the state variables r and λ, and the constant terms:

$$\begin{aligned} &[A_t + \kappa_r A + \kappa_\lambda B + 1] \cdot r \\ &+ [B_t + \theta_\lambda B - \sigma_r A] \cdot \lambda \\ &+ \left[\tfrac{1}{2}\sigma_r^2 A^2 - \mu_r A + \tfrac{1}{2}\sigma_\lambda^2 B^2 - (\mu_\lambda - \sigma_\lambda \lambda_2) B + \rho\sigma_r\sigma_\lambda AB - C_t\right] = 0. \end{aligned} \tag{10.2}$$

In order to hold for every values of $r(t)$ and $\lambda(t)$, we need to have a value of zero for the expressions in brackets. This leads us to a system of decoupled partial differential equations in $A(t)$, $B(t)$, and $C(t)$:

$$A_t + \kappa_r A + \kappa_\lambda B = -1 \tag{10.3}$$

$$B_t + \theta_\lambda B - \sigma_r A = 0 \tag{10.4}$$

$$\frac{1}{2}\sigma_r^2 A^2 - \mu_r A + \frac{1}{2}\sigma_\lambda^2 B^2 - (\mu_\lambda - \sigma_\lambda \lambda_2) B + \rho\sigma_r\sigma_\lambda AB - C_t = 0. \tag{10.5}$$

Looking closer at the system, we fortunately see that function C only appears in the last equation (10.5) and only in its first derivative with respect to t. Thus, we can start solving the system by examining the first two equations

(10.3) and (10.4) in order to derive a solution for the functions $A(t)$ and $B(t)$. In matrix notation we have

$$\begin{bmatrix} A_t \\ B_t \end{bmatrix} + \left[\mathbf{T}^{\mathbb{Q}}\right]' \begin{bmatrix} A \\ B \end{bmatrix} = \mathbf{d},$$

with the *system matrix* $\left[\mathbf{T}^{\mathbb{Q}}\right]'$ and the *driving term* $\mathbf{d} = [-1 \;\; 0]'$. The general solution to this non-homogeneous subsystem is the sum of a particular solution to the non-homogeneous equation (denoted by A^p, and B^p) and the general solution to the homogeneous equation (complementary functions A^c, and B^c):

$$\begin{aligned} A(t) &= A^p + A^c, \text{ and} \\ B(t) &= B^p + B^c. \end{aligned}$$

For the particular solution we can gain the constant solution by solving the following system:

$$\left[\mathbf{T}^{\mathbb{Q}}\right]' \begin{bmatrix} A^p \\ B^p \end{bmatrix} = \mathbf{d}.$$

Thus, we obtain the functions

$$A^p = \frac{\begin{vmatrix} -1 & \kappa_\lambda \\ 0 & \theta_\lambda \end{vmatrix}}{\begin{vmatrix} \kappa_r & \kappa_\lambda \\ -\sigma_r & \theta_\lambda \end{vmatrix}} = \frac{-\theta_\lambda}{\kappa_r\theta_\lambda + \kappa_\lambda\sigma_r}, \text{ and } B^p = \frac{\begin{vmatrix} \kappa_r & -1 \\ -\sigma_r & 0 \end{vmatrix}}{\begin{vmatrix} \kappa_r & \kappa_\lambda \\ -\sigma_r & \theta_\lambda \end{vmatrix}} = \frac{-\sigma_r}{\kappa_r\theta_\lambda + \kappa_\lambda\sigma_r}$$

for the particular solution. Now, we look at the general solutions A^c and B^c to the homogeneous equations

$$\begin{bmatrix} A_t^c \\ B_t^c \end{bmatrix} + \left[\mathbf{T}^{\mathbb{Q}}\right]' \begin{bmatrix} A^c \\ B^c \end{bmatrix} = \mathbf{0}. \tag{10.6}$$

Here, we adopt the trial solution

$$A^c = \alpha e^{\Lambda t}, \text{ and } B^c = \beta e^{\Lambda t}.$$

Putting these with the corresponding time derivatives into equation (10.6), we get

$$\begin{aligned} \Lambda\alpha e^{\Lambda t} + \kappa_r \alpha e^{\Lambda t} + \kappa_\lambda \beta e^{\Lambda t} &= 0 \\ \Lambda\beta e^{\Lambda t} - \sigma_r \alpha e^{\Lambda t} + \theta_\lambda \beta e^{\Lambda t} &= 0 \end{aligned}$$

and

$$(\kappa_r + \Lambda) \cdot \alpha + \kappa_\lambda \cdot \beta = 0 \tag{10.7}$$
$$-\sigma_r \cdot \alpha + (\theta_\lambda + \Lambda) \cdot \beta = 0. \tag{10.8}$$

We note that equations (10.7) and (10.8) only have non-trivial solutions, if the *characteristic polynomial* is equal to zero, i.e.

$$\begin{vmatrix} \kappa_r + \Lambda & \kappa_\lambda \\ -\sigma_r & \theta_\lambda + \Lambda \end{vmatrix} = 0,$$

or

$$\Lambda^2 + (\kappa_r + \theta_\lambda)\,\Lambda + \kappa_r\theta_\lambda + \kappa_\lambda\sigma_r = 0. \tag{10.9}$$

This results in $\Lambda_i = -\frac{1}{2}(\kappa_r + \theta_\lambda) \pm \frac{1}{2}\sqrt{(\kappa_r + \theta_\lambda)^2 - 4(\kappa_r\theta_\lambda + \kappa_\lambda\sigma_r)}$ for the eigenvalues, with $i \in \{1, 2\}$.

Remark 10.2 (Special Eigenvalues) *We need to note the special cases of the eigenvalues where the following calculations are not defined: (i) $\Lambda_i = 0$, and (ii) $\Lambda_1 = \Lambda_2$.*[1] *However, if the parameters are estimated empirically, there seems to be a neglectable possibility of ending up with this special cases.*

Now, we can resolve the values for α and β from, for example, equation (10.8) to $\alpha/\beta = (\theta_\lambda + \Lambda)/\sigma_r$. Choosing arbitrary constants k_i, with $i \in \{1, 2\}$, we get $\alpha_i = (\theta_\lambda + \Lambda_i) k_i$, and $\beta_i = \sigma_r k_i$. Thus we get for the complementary functions

$$A^c = (\theta_\lambda + \Lambda_1) k_1 e^{\Lambda_1 t} + (\theta_\lambda + \Lambda_2) k_2 e^{\Lambda_2 t}, \text{ and}$$
$$B^c = \sigma_r k_1 e^{\Lambda_1 t} + \sigma_r k_2 e^{\Lambda_2 t}.$$

For an intermediate result, we now combine the particular and the complementary solution for the functions $A(t)$ and $B(t)$. Making use of Vieta's Theorem, by which the product of the eigenvalues Λ_i, with $i \in \{1, 2\}$, equals the constant term $\kappa_r\theta_\lambda + \kappa_\lambda\sigma_r$ in equation (10.9)[2], we obtain:

$$A(t) = -\frac{\theta_\lambda}{\Lambda_1\Lambda_2} + (\theta_\lambda + \Lambda_1) k_1 e^{\Lambda_1 t} + (\theta_\lambda + \Lambda_2) k_2 e^{\Lambda_2 t}, \tag{10.10}$$
$$B(t) = -\frac{\sigma_r}{\Lambda_1\Lambda_2} + \sigma_r k_1 e^{\Lambda_1 t} + \sigma_r k_2 e^{\Lambda_2 t}. \tag{10.11}$$

[1] As we discuss in section 10.3.1, to get a solution for $\Lambda_i = 0$ we need to go back to the original system of partial differential equations of (10.3) to (10.5).

[2] This equivalently is the determinant of the system matrix.

In order to derive the constants k_1 and k_2, we now make use of the first two boundary conditions $A(T) = 0$ and $B(T) = 0$. For equations (10.10) and (10.11) we get in matrix form:

$$\begin{bmatrix} \theta_\lambda + \Lambda_1 & \theta_\lambda + \Lambda_2 \\ 1 & 1 \end{bmatrix} \begin{bmatrix} k_1 e^{\Lambda_1 T} \\ k_2 e^{\Lambda_2 T} \end{bmatrix} = \begin{bmatrix} \frac{\theta_\lambda}{\Lambda_1 \Lambda_2} \\ \frac{1}{\Lambda_1 \Lambda_2} \end{bmatrix},$$

with the solutions

$$k_1 e^{\Lambda_1 T} = \frac{\begin{vmatrix} \frac{\theta_\lambda}{\Lambda_1 \Lambda_2} & \theta_\lambda + \Lambda_2 \\ \frac{1}{\Lambda_1 \Lambda_2} & 1 \end{vmatrix}}{\begin{vmatrix} \theta_\lambda + \Lambda_1 & \theta_\lambda + \Lambda_2 \\ 1 & 1 \end{vmatrix}} = \frac{-1}{\Lambda_1 (\Lambda_1 - \Lambda_2)}$$

$$k_1 = \frac{-1}{\Lambda_1 (\Lambda_1 - \Lambda_2)} e^{-\Lambda_1 T}, \text{ and}$$

$$k_2 e^{\Lambda_2 T} = \frac{\begin{vmatrix} \theta_\lambda + \Lambda_1 & \frac{\theta_\lambda}{\Lambda_1 \Lambda_2} \\ 1 & \frac{1}{\Lambda_1 \Lambda_2} \end{vmatrix}}{\begin{vmatrix} \theta_\lambda + \Lambda_1 & \theta_\lambda + \Lambda_2 \\ 1 & 1 \end{vmatrix}} = \frac{1}{\Lambda_2 (\Lambda_1 - \Lambda_2)}$$

$$k_2 = \frac{1}{\Lambda_2 (\Lambda_1 - \Lambda_2)} e^{-\Lambda_2 T}.$$

This finally results in the values for the functions $A(t,T)$ and $B(t,T)$:[3]

$$A(t,T) = a_0 + a_1 e^{-\Lambda_1 (T-t)} + a_2 e^{-\Lambda_2 (T-t)}, \tag{10.12}$$

$$\text{with } a_0 = -\frac{\theta_\lambda}{\Lambda_1 \Lambda_2}, \quad a_1 = -\frac{\theta_\lambda + \Lambda_1}{\Lambda_1 (\Lambda_1 - \Lambda_2)}, \quad a_2 = \frac{\theta_\lambda + \Lambda_2}{\Lambda_2 (\Lambda_1 - \Lambda_2)}, \quad \text{and}$$

$$B(t,T) = b_0 + b_1 e^{-\Lambda_1 (T-t)} + b_2 e^{-\Lambda_2 (T-t)}, \tag{10.13}$$

$$\text{with } b_0 = -\frac{\sigma_r}{\Lambda_1 \Lambda_2}, \quad b_1 = -\frac{\sigma_r}{\Lambda_1 (\Lambda_1 - \Lambda_2)}, \quad b_2 = \frac{\sigma_r}{\Lambda_2 (\Lambda_1 - \Lambda_2)}.$$

With these results for $A(t,T)$ and $B(t,T)$, we can now solve equation (10.5) for the last function of relevance $C(t)$.

[3] For a better understanding of the independent variables of the functions $A(t)$ and $B(t)$ we add the parameter T to the final expressions, i.e. denote the functions as $A(t,T)$ and $B(t,T)$.

Remark 10.3 (Alternative Derivation) *Instead of solving the system of differential equations consisting of (10.3) and (10.4), we alternatively can derive the solutions of $A(t)$ and $B(t)$ by reducing the two first-order differential equations to a single second-order equation. By differentiating equation (10.3) with respect to t, and inserting it into equation (10.4), we can defer the following single second-order equation for $A(t)$:*

$$A_{tt} + (\kappa_r + \theta_\lambda) \cdot A_t + (\kappa_r \theta_\lambda + \kappa_\lambda \sigma_r) \cdot A = -\theta_\lambda.$$

Having derived a solution for $A(t)$, we then resolve equation (10.4) for the function $B(t)$. With the boundary conditions $A(T) = 0$ and $B(T) = 0$, we finally get the same results for the functions $A(t)$, and $B(t)$.[4]

Finally, we start calculating $C(t)$ by isolating the partial derivative C_t in the third equation of our system (10.5):

$$C_t = \frac{1}{2}\sigma_r^2 A^2 - \mu_r A + \frac{1}{2}\sigma_\lambda^2 B^2 - (\mu_\lambda - \sigma_\lambda \lambda_2) B + \rho \sigma_r \sigma_\lambda AB,$$

where we now enter our results of equations (10.12) and (10.13) for $A(t)$ and $B(t)$. A straightforward calculation gives us in a nice rearrangement with constants c_i, $i \in \{0, 1, ..., 5\}$:

$$\begin{aligned} C_t &= c_0 + c_1 e^{-\Lambda_1(T-t)} + c_2 e^{-\Lambda_2(T-t)} \\ &\quad + c_3 e^{-2\Lambda_1(T-t)} + c_4 e^{-2\Lambda_2(T-t)} + c_5 e^{-(\Lambda_1+\Lambda_2)(T-t)} \end{aligned} \tag{10.14}$$

$$\begin{aligned} \text{with } c_0 &= \frac{1}{2}\sigma_r^2 a_0^2 - \mu_r a_0 + \frac{1}{2}\sigma_\lambda^2 b_0^2 - (\mu_\lambda - \sigma_\lambda \lambda_2) b_0 + \rho \sigma_r \sigma_\lambda a_0 b_0 \\ c_1 &= \left(\sigma_r^2 a_0 - \mu_r + \rho \sigma_r \sigma_\lambda b_0\right) a_1 + \left(\sigma_\lambda^2 b_0 - (\mu_\lambda - \sigma_\lambda \lambda_2) + \rho \sigma_r \sigma_\lambda a_0\right) b_1 \\ c_2 &= \left(\sigma_r^2 a_0 - \mu_r + \rho \sigma_r \sigma_\lambda b_0\right) a_2 + \left(\sigma_\lambda^2 b_0 - (\mu_\lambda - \sigma_\lambda \lambda_2) + \rho \sigma_r \sigma_\lambda a_0\right) b_2 \\ c_3 &= \frac{1}{2}\sigma_r^2 a_1^2 + \rho \sigma_r \sigma_\lambda a_1 b_1 + \frac{1}{2}\sigma_\lambda^2 b_1^2 \\ c_4 &= \frac{1}{2}\sigma_r^2 a_2^2 + \rho \sigma_r \sigma_\lambda a_2 b_2 + \frac{1}{2}\sigma_\lambda^2 b_2^2 \\ c_5 &= \sigma_r^2 a_1 a_2 + \rho \sigma_r \sigma_\lambda (a_1 b_2 + a_2 b_1) + \sigma_\lambda^2 b_1 b_2. \end{aligned}$$

[4] This alternative derivation is of interest in terms of studying the equilibrium behavior of the system, because it nicely relates to the nonhomogeneous partial differential equation which is known from physics to model oscillation. However, our approach is more symmetric in its formulation and derivation of the solutions to $A(t)$ and $B(t)$.

Finally, in view of the boundary condition $C(T) = 0$, we integrate equation (10.14) to derive the result for the function $C(t)$

$$-C(t) = \int_t^T c_0 + c_1 e^{-\Lambda_1(T-\zeta)} + c_2 e^{-\Lambda_2(T-\zeta)} + c_3 e^{-2\Lambda_1(T-\zeta)} + c_4 e^{-2\Lambda_2(T-\zeta)} + c_5 e^{-(\Lambda_1+\Lambda_2)(T-\zeta)} d\zeta,$$

which yields:[5]

$$\begin{aligned} C(t,T) = & -c_0(T-t) + \frac{c_1}{\Lambda_1}\left(e^{-\Lambda_1(T-t)} - 1\right) + \frac{c_2}{\Lambda_2}\left(e^{-\Lambda_2(T-t)} - 1\right) \\ & + \frac{c_3}{2\Lambda_1}\left(e^{-2\Lambda_1(T-t)} - 1\right) + \frac{c_4}{2\Lambda_2}\left(e^{-2\Lambda_2(T-t)} - 1\right) \\ & + \frac{c_5}{\Lambda_1 + \Lambda_2}\left(e^{-(\Lambda_1+\Lambda_2)(T-t)} - 1\right), \end{aligned} \tag{10.15}$$

where the constants c_i, with $i \in \{0, 1, ..., 5\}$, are given in equation (10.14).

Solution 10.4 (Discount Bond Price) *Putting our results together, we are able to state the price of a discount bond in the closed-form expression*

$$P(t,T,\mathbf{x}) = e^{-[A(t,T),B(t,T)]\mathbf{x} - C(t,T)} \tag{10.16}$$

where we use the formulas for the functions $A(t,T)$, $B(t,T)$ and $C(t,T)$ from equations (10.12), (10.13), and (10.15), respectively.

Moreover, with the explicit formulas of the functions $A(t,T)$, $B(t,T)$ and $C(t,T)$ we can readily calculate the risk-neutral dynamics of the discount bond price. We begin with the P-dynamics under the martingale measure $\mathbb{Q}$ of equation (9.3) which generates the stochastic form:

$$\begin{aligned} \frac{dP}{P} = & \left[(-A_t - \kappa_r A - \kappa_\lambda B) r + (-B_t - \theta_\lambda B + \sigma_r A) \lambda \right. \\ & \left. + \frac{1}{2}\sigma_r^2 A^2 - \mu_r A + \frac{1}{2}\sigma_\lambda^2 B^2 - (\mu_\lambda - \sigma_\lambda \lambda_2) B + \rho \sigma_r \sigma_\lambda AB - C_t \right] dt \\ & - \sigma_r A dW_{r,t}^{\mathbb{Q}} - \sigma_\lambda B dW_{\lambda,t}^{\mathbb{Q}}. \end{aligned}$$

In conjunction with the results for $A(t,T)$, $B(t,T)$, and $C(t,T)$ the stochastic dynamics for the discount bond simplifies to

---

5 We also add the parameter T to the final expression on the function $C(t)$ for a better understanding, i.e. we denote the function as $C(t,T)$.

$$dP = \mu_P P dt + \eta_r P dW^{\mathbb{Q}}_{r,t} + \eta_\lambda P dW^{\mathbb{Q}}_{\lambda,t}, \tag{10.17}$$
$$\text{with } \mu_P = r, \;\; \eta_r = -\sigma_r A, \text{ and } \eta_\lambda = -\sigma_\lambda B.$$

Equation (10.17) shows explicitly that the modeled discount bond $P(t, T, \mathbf{x})$ actually has a drift rate of r in the risk neutralized world.

10.2 Term Structures of Interest Rates and Volatilities

On the basis of the main result in the previous section - the formula for the value of a discount bond - we are now able to explore the cross-sectional dimensions of our interest rate model in detail. We examine both the term structure of interest rates as well as the term structure of volatilities in the following.

10.2.1 Spot and Forward Rate Curves

From the general functional relationship between the value of discount bonds and spot rates in section 8.2 we accomplish to determine the yield curve implied by our term structure model. We conveniently compute the yield to maturities $y(t, T)$ of different discount bond prices by the *affine relationship* between the zero bond yield $y(t, T)$ and the state variables $r(t)$ and $\lambda(t)$ with the relationship from equation (8.1)

$$y(t, T) = \frac{1}{T - t}\left[\left[A(t, T), B(t, T)\right]\mathbf{x} + C(t, T)\right]; \tag{10.18}$$

the factor loadings $A(t, T)$ and $B(t, T)$ are given in equations (10.12) and (10.13), and the intercept term $C(t, T)$ is from equation (10.15). Especially this affine structure of the yields in the driving factors - the short rate r and the market price of risk $\lambda(t)$ - will be of special importance for the empirical implementation and analysis which we perform in chapter 15.

Further, it is of interest how the forward rate curve looks in this model. Starting from equation (8.2) we can see that forward rates are related to spot rates by

$$f(t, T_1, T_2) = \frac{y(t, T_2)(T_2 - t) - y(t, T_1)(T_1 - t)}{T_2 - T_1}. \tag{10.19}$$

Thus, the term structure of forward rates can immediately be derived from the zero bond yield curve given in equation (10.18).

10.2.2 Term Structure of Volatilities

Besides the well known meaning and usage of the phrase *term structure* as the term structure of interest rates - as in the previous section - it is also used in the context of volatilities. The term describes the functional relationship of either the volatilities of the spot or forward rates over the spectrum of different time to maturities. This second meaning probably stems from the importance of volatilities in option pricing as we will see further down in chapter 14.

In order to derive the volatility structure of the *spot rates* we first need to determine the dynamics of the T-maturity spot rate $y(t,T)$. In our specification of the term structure with the risk-neutral factor processes given in equation (9.2), we attain the following stochastic differential on $y(t,T)$ by applying Itô's formula:

$$dy(t,T) = \mu_y dt + \sigma_r y_r dW^{\mathbb{Q}}_{r,t} + \sigma_\lambda y_\lambda dW^{\mathbb{Q}}_{\lambda,t},$$
$$\text{with } \mu_y = \mu_1 y_r + \mu_2 y_\lambda + y_t + \frac{1}{2}\sigma_r^2 y_{rr} + \rho\sigma_r\sigma_\lambda y_{r\lambda} + \frac{1}{2}\sigma_\lambda^2 y_{\lambda\lambda}.$$

Therein, the partial derivatives y_r, y_λ, y_t, y_{rr}, $y_{r\lambda}$, and $y_{\lambda\lambda}$ are easily found from the result of equation (10.18). Using these derivatives leads us to

$$dy(t,T) = \left(\mu_1 \frac{A(t,T)}{T-t} + \mu_2 \frac{B(t,T)}{T-t}\right) dt$$
$$+\sigma_r \frac{A(t,T)}{T-t} dW^{\mathbb{Q}}_{r,t} + \sigma_\lambda \frac{B(t,T)}{T-t} dW^{\mathbb{Q}}_{\lambda,t}. \tag{10.20}$$

To further calculate the volatility structure of spot rates, we use the integrated form of equation (10.20) over the period $[t, t+\Delta t]$ to yield

$$y(t+\Delta t, T+\Delta t) = y(t,T) + \left(\mu_1 \frac{A(t,T)}{T-t} + \mu_2 \frac{B(t,T)}{T-t}\right) \Delta t$$
$$+ \int_t^{t+\Delta t} \sigma_r \frac{A(t,T)}{T-t} dW^{\mathbb{Q}}_{r,\varsigma} + \int_t^{t+\Delta t} \sigma_\lambda \frac{B(t,T)}{T-t} dW^{\mathbb{Q}}_{\lambda,\varsigma},$$

where the two stochastic integrals follow a normal distribution. Thus, the spot rate $y(t,T)$ follows a bivariate Gaussian distribution. From this dynamics we obtain the variance per unit of time, i.e. $\Delta t = 1$, as:

$$\mathbb{V}ar\left[y(t,T)\right] = \mathbb{E}^{\mathbb{Q}}\left[y_t^2 | \mathcal{F}_s\right] - \mathbb{E}^{\mathbb{Q}}\left[y_t | \mathcal{F}_s\right]^2$$
$$= \left(\sigma_r \frac{A(t,T)}{T-t}\right)^2 + \left(\sigma_\lambda \frac{B(t,T)}{T-t}\right)^2$$
$$+2\rho\sigma_r\sigma_\lambda \frac{A(t,T)B(t,T)}{(T-t)^2}, \tag{10.21}$$

which describes the *term structure of spot rate volatilities.*

One important capability of a two-factor model is its implication of an imperfect correlation of spot rates of different maturities. Therefore we calculate the covariance between the interest rates of discount bonds with different maturities:

$$\begin{aligned}\mathbb{E}^{\mathbb{Q}}\left[dy\left(t,T_1\right)dy\left(t,T_2\right)\right] &= \mathbb{E}^{\mathbb{Q}}\left[\frac{\sigma_r A\left(t,T_1\right)dW_{r,t}^{\mathbb{Q}}+\sigma_\lambda B\left(t,T_1\right)dW_{\lambda,t}^{\mathbb{Q}}}{T_1-t}\right.\\ &\qquad\left.\times\frac{\sigma_r A\left(t,T_2\right)dW_{r,t}^{\mathbb{Q}}+\sigma_\lambda B\left(t,T_2\right)dW_{\lambda,t}^{\mathbb{Q}}}{T_2-t}\right]\\ &= \left[\frac{\sigma_r^2 A\left(t,T_1\right)A\left(t,T_2\right)+\sigma_\lambda^2 B\left(t,T_1\right)B\left(t,T_2\right)}{\left(T_1-t\right)\left(T_2-t\right)}\right]dt\\ &\quad+\left[\frac{A\left(t,T_1\right)B\left(t,T_2\right)+A\left(t,T_2\right)B\left(t,T_1\right)}{\left(T_1-t\right)\left(T_2-t\right)}\right]\\ &\quad\rho\sigma_r\sigma_\lambda dt.\end{aligned}$$

We further use the general relationship for the covariance

$$\mathbb{E}^{\mathbb{Q}}\left[dy\left(t,T_1\right)dy\left(t,T_2\right)\right]=\rho_{T_1T_2}\sigma_{T_1}\sigma_{T_2}dt$$

to reveal our final result for the correlation coefficient of the spot rates with maturities T_1-t and T_2-t:

$$\begin{aligned}\rho_{T_1T_2} &= \frac{1}{\sigma_{T_1}\sigma_{T_2}\left(T_1-t\right)\left(T_2-t\right)}\left[\sigma_r^2 A\left(t,T_1\right)A\left(t,T_2\right)+\sigma_\lambda^2 B\left(t,T_1\right)\right.\\ &\quad\left. B\left(t,T_2\right)+\rho\sigma_r\sigma_\lambda\left(A\left(t,T_1\right)B\left(t,T_2\right)+A\left(t,T_2\right)B\left(t,T_1\right)\right)\right].\end{aligned}\tag{10.22}$$

In analogy to the previous findings we are able to derive the volatility structure on the *forward rates.* From the result of equation (10.19) we obtain the stochastic dynamics

$$\begin{aligned}df\left(t,T\right) &= \left(\mu_1\frac{A\left(t,T\right)}{T-t}+\mu_2\frac{B\left(t,T\right)}{T-t}\right)dt\\ &\quad+\sigma_r\frac{\partial A\left(t,T\right)}{\partial T}dW_{r,t}^{\mathbb{Q}}+\sigma_\lambda\frac{\partial B\left(t,T\right)}{\partial T}dW_{\lambda,t}^{\mathbb{Q}}\end{aligned}$$

for the instantaneous forward rate. By using the integrated form of this dynamics as an intermediate step, we attain the *term structure of forward rate volatilities* as:

$$\begin{aligned}\mathbb{V}ar\left[f\left(t,T\right)\right] &= \left(\frac{\sigma_r}{T-t}\frac{\partial A\left(t,T\right)}{\partial T}\right)^2+\left(\frac{\sigma_\lambda}{T-t}\frac{\partial B\left(t,T\right)}{\partial T}\right)^2\\ &\quad+\frac{2\rho\sigma_r\sigma_\lambda}{\left(T-t\right)^2}\frac{\partial A\left(t,T\right)}{\partial T}\frac{\partial B\left(t,T\right)}{\partial T}.\end{aligned}$$

We finally calculate the covariance between forward rates of different maturities $T_1 - t$ and $T_2 - t$ as

$$\begin{aligned}\mathbb{C}ov\left[f\left(t,T_1\right),f\left(t,T_2\right)\right] = \frac{1}{\left(T_1-t\right)\left(T_2-t\right)}&\left[\sigma_r^2\frac{\partial A\left(t,T_1\right)}{\partial T}\frac{\partial A\left(t,T_2\right)}{\partial T}\right.\\&+\sigma_\lambda^2\frac{\partial B\left(t,T_1\right)}{\partial T}\frac{\partial B\left(t,T_2\right)}{\partial T}\\&+\rho\sigma_r\sigma_\lambda\left(\frac{\partial A\left(t,T_1\right)}{\partial T}\frac{\partial B\left(t,T_2\right)}{\partial T}\right.\\&\left.\left.+\frac{\partial A\left(t,T_2\right)}{\partial T}\frac{\partial B\left(t,T_1\right)}{\partial T}\right)\right]\end{aligned}$$

to complete the analysis on the term structure of volatilities.

10.3 Analysis of Limiting Cases

10.3.1 Reducing to an Ornstein-Uhlenbeck Process

Since our model is an extended version of modeling the short rate dynamics with an Ornstein-Uhlenbeck process, we would naturally think of comparing our pricing formula in equation (10.16) to that of Vasicek (1977). However, the derived pricing formula is not valid in the case where we set the time homogeneous parameters μ_λ, θ_λ, σ_λ, and λ_2 equal to zero. Instead, as noted in remark 10.2, we already need to adjust the parameters to their appropriate values in the system of partial differential equations in (10.3), (10.4), and (10.5). This reduces the system of equations to

$$A_t + \kappa_r A = -1,\ B_t - \sigma_r A = 0,\text{ and } \frac{1}{2}\sigma_r^2 A^2 - \mu_r A - C_t = 0. \qquad (10.23)$$

In solving this reduced system, we use the same boundary conditions as in proposition 10.1 for the affine solution $P(t,T,\mathbf{x}) = e^{-[A(t),B(t)]\mathbf{x}-C(t)}$ where, however, the market price of interest rate risk is now a constant, i.e. $\lambda = const.$ First, we calculate the solution to the function $A(t)$ out of the first equation in (10.23). A particular solution is easily seen to be $A^p = -1/\kappa_r$; we further use the homogeneous equation to calculate the complementary solution with integration constants k and k':

$$\begin{aligned}\int_t^T \frac{1}{A^c}dA^c &= -\kappa_r\int_t^T d\zeta\\ -\ln A^c(t) &= -\kappa_r(t-T)+k\\ A^c(t) &= k'e^{\kappa_r(T-t)}.\end{aligned}$$

Now, we add the particular and complementary solution, and incorporate the boundary condition $A(T) = 0$ to find the general solution to be

$$A(t) = -\frac{1}{\kappa_r}\left(1 - e^{\kappa_r(T-t)}\right). \tag{10.24}$$

For the function $B(t)$, we get from the second homogeneous equation in (10.23) by using $B(T) = 0$ the following result:

$$\begin{aligned}
\int_t^T dB &= -\frac{\sigma_r}{\kappa_r}\int_t^T \left(1 - e^{\kappa_r(T-\zeta)}\right) d\zeta \\
-B(t) &= -\frac{\sigma_r}{\kappa_r}\left((T-t) + \frac{1}{\kappa_r}\left(1 - e^{\kappa_r(T-t)}\right)\right) \\
B(t) &= -\frac{\sigma_r}{\kappa_r}\left[A(t) - (T-t)\right].
\end{aligned} \tag{10.25}$$

In order to derive the function $C(t)$ with its boundary condition $C(T) = 0$, we use the last equation of equations (10.23):

$$\begin{aligned}
C_t &= \frac{\sigma_r^2}{2\kappa_r^2}\left(1 - e^{\kappa_r(T-t)}\right)^2 + \frac{\mu_r}{\kappa_r}\left(1 - e^{\kappa_r(T-t)}\right) \\
\int_t^T dC &= \int_t^T \frac{\sigma_r^2}{2\kappa_r^2} + \frac{\mu_r}{\kappa_r} - \left(\frac{\sigma_r^2}{\kappa_r} + \frac{\mu_r}{\kappa_r}\right) e^{\kappa_r(T-\zeta)} + \frac{\sigma_r^2}{2\kappa_r} e^{2\kappa_r(T-\zeta)} d\zeta \\
-C(t) &= \left(\frac{\sigma_r^2}{2\kappa_r^2} + \frac{\mu_r}{\kappa_r}\right)(T-t) + \left(\frac{\sigma_r^2}{\kappa_r} + \frac{\mu_r}{\kappa_r}\right)\left(1 - e^{\kappa_r(T-t)}\right) \\
&\quad - \frac{\sigma_r^2}{4\kappa_r^3}\left(1 - e^{2\kappa_r(T-t)}\right) \\
C(t) &= \left(-\frac{\mu_r}{\kappa_r} - \frac{\sigma_r^2}{2\kappa_r^2}\right)\left((T-t) - A(t)\right) - \frac{\sigma_r^2}{4\kappa_r}A^2(t).
\end{aligned} \tag{10.26}$$

To see the equivalence between the bond price formula from Vasicek (1977) and our solution

$$P(t, T, \mathbf{x}) = e^{\left[\frac{1}{\kappa_r}\left(1-e^{\kappa_r(T-t)}\right), \frac{\sigma_r}{\kappa_r}\left[A(t)-(T-t)\right]\right]\mathbf{x} + \left(\frac{\mu_r}{\kappa_r} + \frac{\sigma_r^2}{2\kappa_r^2}\right)\left((T-t)-A(t)\right) + \frac{\sigma_r^2}{4\kappa_r}A^2(t)},$$

we conveniently change the symbols for the used parameters. The original notation of the Ornstein-Uhlenbeck process used by Vasicek (1977, p. 185) is

$$dr = \alpha(\gamma - r)\,dt + dz.$$

With the substitutions $-\kappa_r = \alpha$ and $-\mu_r/\kappa_r = \gamma$ we are easily able to transfer our results into that found by Vasicek (1977).

10.3.2 Examining the Asymptotic Behavior

When looking at possible shapes of the term structure of interest rates two points of the term structure are of special interest: The short end and the long end. These are given by two extreme values for the maturity parameter T. We examine the behavior of the yield function from equation (10.18) for the case when T approximates t from above, and for the case where T approaches infinity.

For theoretical reasons the *short end* of the term structure should coincide with the instantaneous rate of return $r(t)$. This implies that the functions $A(t)$, $B(t)$, and $C(t)$ should have limit values of

$$\lim_{T\downarrow t}\frac{A(t)}{T-t}=1,\text{ and }\lim_{T\downarrow t}\frac{B(t)}{T-t}=\lim_{T\downarrow t}\frac{C(t)}{T-t}=0.$$

For calculating the limit values we use the method of Taylor series expansion. In our case a first order approximation is sufficient. For the limit values of the functions $A(t)$, $B(t)$, and $C(t)$ we derive:

$$\begin{aligned}\lim_{T\downarrow t}\frac{A(t)}{T-t} &= \frac{1}{T-t}\left[a_0+a_1\left(1-\Lambda_1(T-t)+o(T-t)\right)\right.\\ &\quad\left.+a_2\left(1-\Lambda_2(T-t)+o(T-t)\right)\right]\\ &= \frac{1}{T-t}\left[\frac{-\theta_\lambda(\Lambda_1-\Lambda_2)-(\theta_\lambda+\Lambda_1)\Lambda_2+(\theta_\lambda+\Lambda_2)\Lambda_1}{\Lambda_1\Lambda_2(\Lambda_1-\Lambda_2)}\right.\\ &\quad\left.+\frac{\theta_\lambda+\Lambda_1-(\theta_\lambda+\Lambda_2)}{\Lambda_1-\Lambda_2}(T-t)\right]=1,\text{ and}\end{aligned}$$

$$\begin{aligned}\lim_{T\downarrow t}\frac{B(t)}{T-t} &= \frac{1}{T-t}\left[b_0+b_1\left(1-\Lambda_1(T-t)+o(T-t)\right)\right.\\ &\quad\left.+b_2\left(1-\Lambda_2(T-t)+o(T-t)\right)\right]\\ &= \frac{1}{T-t}\left[\frac{-\sigma_r(\Lambda_1-\Lambda_2)-\sigma_r\Lambda_2+\sigma_r\Lambda_1}{\Lambda_1\Lambda_2(\Lambda_1-\Lambda_2)}\right.\\ &\quad\left.+\frac{\sigma_r-\sigma_r}{\Lambda_1-\Lambda_2}(T-t)\right]=0,\text{ and}\end{aligned}$$

$$
\begin{aligned}
\lim_{T\downarrow t}\frac{C(t)}{T-t} &= \frac{1}{T-t}\left[-c_0(T-t)+\frac{c_1}{\Lambda_1}\left(1-\Lambda_1(T-t)+o(T-t)-1\right)\right.\\
&\quad+\frac{c_2}{\Lambda_2}\left(1-\Lambda_2(T-t)+o(T-t)-1\right)\\
&\quad+\frac{c_3}{2\Lambda_1}\left(1-\Lambda_1(T-t)+o(T-t)-1\right)\\
&\quad+\frac{c_4}{2\Lambda_2}\left(1-2\Lambda_2(T-t)+o(T-t)-1\right)\\
&\quad\left.+\frac{c_5}{\Lambda_1+\Lambda_2}\left(1-(\Lambda_1+\Lambda_2)(T-t)+o(T-t)-1\right)\right]\\
&= -\left[c_0+c_1+c_2+c_3+c_4+c_5\right]=0.
\end{aligned}
$$

By having calculated these limiting values for the functions $A(t)$, $B(t)$, and $C(t)$ we indirectly verified our pricing formula of equation (10.16) in that we obtain

$$
y_s(t)=\lim_{T\downarrow t}\frac{y(t,T)}{T-t}=\lim_{T\downarrow t}\frac{1}{T-t}\left[\left[A(t),B(t)\right]\mathbf{x}+C(t)\right]=r(t)
$$

for the short end of the term structure of interest rates. Thus the short end coincides with the first state variable of our term structure model.

Now, we analyze the asymptotic behavior of the term structure of interest rates at its *long end*. For the long end we need to investigate the limiting value

$$
\lim_{T\to\infty}y(t,T)=\lim_{T\to\infty}\frac{1}{T-t}\left[\left[A(t)\quad B(t)\right]\mathbf{x}+C(t)\right].
$$

Again, we examine the three functions $A(t)$, $B(t)$, and $C(t)$ separately:

$$
\begin{aligned}
\lim_{T\to\infty}\frac{A(t)}{T-t} &= \lim_{T\to\infty}\frac{1}{T-t}\left[a_0+a_1e^{-\Lambda_1(T-t)}+a_2e^{-\Lambda_2(T-t)}\right]=0,\text{ and}\\
\lim_{T\to\infty}\frac{B(t)}{T-t} &= \lim_{T\to\infty}\frac{1}{T-t}\left[b_0+b_1e^{-\Lambda_1(T-t)}+b_2e^{-\Lambda_2(T-t)}\right]=0,\text{ and}\\
\lim_{T\to\infty}\frac{C(t)}{T-t} &= \lim_{T\to\infty}\frac{1}{T-t}\left[-c_0(T-t)+\frac{c_1}{\Lambda_1}\left(e^{-\Lambda_1(T-t)}-1\right)\right.\\
&\quad+\frac{c_2}{\Lambda_2}\left(e^{-\Lambda_2(T-t)}-1\right)+\frac{c_3}{2\Lambda_1}\left(e^{-2\Lambda_1(T-t)}-1\right)\\
&\quad\left.+\frac{c_4}{2\Lambda_2}\left(e^{-2\Lambda_2(T-t)}-1\right)+\frac{c_5}{\Lambda_1+\Lambda_2}\left(e^{-(\Lambda_1+\Lambda_2)(T-t)}-1\right)\right]\\
&= -c_0.
\end{aligned}
$$

Thus, we obtain the constant for the long end of the term structure

$$
\begin{aligned}
y_l(t) &= \lim_{T\to\infty}y(t,T)\\
&= -\frac{1}{2}\sigma_r^2a_0^2+\mu_ra_0-\frac{1}{2}\sigma_\lambda^2b_0^2+(\mu_\lambda-\sigma_\lambda\lambda_2)\,b_0-\rho\sigma_r\sigma_\lambda a_0b_0,
\end{aligned}
$$

which is independent of the state variables $\mathbf{x} = [r(t), \lambda(t)]'$.

10.4 Possible Shapes of the Term Structures

From the various term structures we derived in section 10.2 we consider the effects on the two special cases of the term structure of spot interest rates as well as the term structure of volatilities for spot interest rates. Our two-factor model endogenously produces these term structures determined by equations (10.18) and (10.21). Both term structures depend on the *market expectations* of the two state variables, as inherently modeled by the specifications of the stochastic processes under the measures $\mathbb{P}$ and $\mathbb{Q}$. These process specifications, respectively, model the dynamics of the state variables through time and determine the term structure relationship between the interest rates of different maturities for a given date. Furthermore, the term structure of spot interest rates is also a function of the *current levels* of the two state variables - the short rate r and the market price of risk λ - which we therefore also need to include in our comparative statistics.

10.4.1 Influences of the State Variables

We first examine the influences of the state variables on the term structure of spot interest rates. As determined by equation (10.18) the term structure is influenced both by the current level of the short rate r_t and the value of the market price of risk λ_t whereas the term structure of volatility is independent of the state variables.[6]

For the exposition of the comparative static analysis we choose to use the empirical estimates for the parameter values ψ reported in table 12.5 as basis.[7] Given these parameter values we work with the specification

$$
\begin{aligned}
dr &= [\mu_r + \kappa_r r - \sigma_r \lambda]\, dt + \sigma_r dW^{\mathbb{Q}}_{r,t} \\
&= [0.0071 - 0.0744r - 0.0225\lambda]\, dt + 0.0225 dW^{\mathbb{Q}}_{r,t}, \text{ and} \qquad (10.27) \\
d\lambda &= [\mu_\lambda - \sigma_\lambda \lambda_2 + \kappa_\lambda r + \theta_\lambda \lambda]\, dt + \sigma_\lambda dW^{\mathbb{Q}}_{\lambda,t} \\
&= [-0.0441 - 0.6198r - 0.4434\lambda]\, dt + 0.3863 dW^{\mathbb{Q}}_{\lambda,t}, \qquad (10.28)
\end{aligned}
$$

[6] See equation (10.21).

[7] The empirical parameter estimates will be considered in depth in chapter 12 on the calibration of the model to US interest rate data.

for the stochastic processes for the state variables including a correlation coefficient of $\rho = 0.7102$ between the two Brownian motions. For the current values of the state variables we use $r_t = 0.0699$ and $\lambda_t = -0.2624$, respectively, which reflect the empirical mean values of the realized state variables.[8]

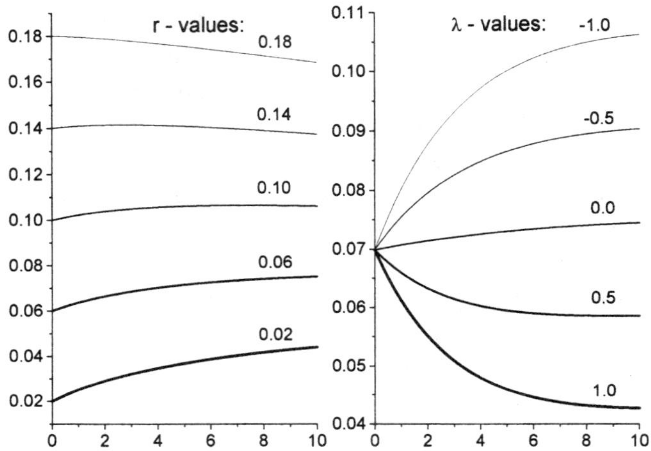

Fig. 10.1. Influence of the State Variables on the Term Structure

How the term structure of spot interest rates reacts to different levels of the state variables is shown in figure 10.1. The left graph is based on a market price of risk of $\lambda_t = -0.2624$ and models the impact of short rates in the range $r_t = \{0.02, 0.04, \ldots, 0.18\}$. The different values for the short rate can be traced back to the crossing points of the yield curves with the ordinate as demonstrated in section 10.3.2. All yield curves are a little upward sloping since the market price of risk ($\lambda_t = -0.2624$) lies below its risk-neutral value of $\lambda_t = 0$. The changes in the short rate have the impact of an approximately parallel shift of the yield curve. The effect is slightly higher on the short end of the yield curve than for longer maturities. Therefore, the first state variable can be attributed as a *level* factor.

The graph on the right side of figure 10.1 underlies a value of $r_t = 0.0699$ for the current short rate. Depicted are yield curves for varying risk attitude values λ_t. To represent the spectrum of risk averse, risk neutral, up to risk taking investors we choose values in the range of $\lambda_t = \{-1.0, -0.5, 0, 0.5, 1.0\}$.

[8] The empirical mean values are those obtained for the US Treasury data set shown in table 12.7 on page 163.

These values are obtained on the basis of empirical observable quantities with estimates on the US term structure.[9] For a constant short rate r_t the different levels of the market price of risk have a significant effect on the slope of the yield curve. For the underlying scenario we obtain normal shapes of the yield curve with negative values for the market price of risk, i.e. risk averse market attitudes towards interest rate risk. This is in accordance with the intuition that more risk averse investors require a higher return on their fixed-income investments. Inverse yield curves on the other hand are attainable for positive market prices of risk. We recognize that the level of the market price of risk is the major factor determining the slope of the yield curve. In the conventional sense the second factor can therefore be considered as the *steepness* factor.

Combining variations in both state variables, while holding the model parameters $\boldsymbol{\psi}$ fixed, results in a rich complexity of different shapes of the yield curve. This flexibility is rather relevant in recalibrating the model from time to time with a close fit to the then prevailing term structures without reestimating the whole range of model parameters $\boldsymbol{\psi}$.

The next subject of analysis is the term structure of volatilities. Since the volatility curve is independent of the levels of the state variables, we now move to the influences of choosing particular model parameters on the term structures.

10.4.2 Choosing the Model Parameters

In this section we establish comparative statistical facts on the behavior of the two term structures for different values of the model parameters. The parameters of the theoretical term structure model include $\boldsymbol{\psi} = \{\mu_r, \kappa_r, \sigma_r, \mu_\lambda, \kappa_\lambda, \sigma_\lambda, \theta_\lambda, \lambda_2, \rho\}$. The range of possible values for the parameters is found on the ground of the empirical results on the US term structure data.[10] The analyzed range includes all empirical values as well as some higher and lower values. The standard parameter values are the same as in the previous section.

We first examine the behavior of the term structure of *spot interest rates.* The possible term structures for the nine parameters are plotted in figure 10.2. The graphs are further based on the values of $r_t = 0.0699$ and $\lambda_t = -0.2624$ for the state variables. Since the r_t-value lies below its long-run mean of $r_l = 0.0950$ and the risk attitude is risk averse the resulting yield curve shapes

[9] See the empirical results shown in table 12.7 on page 163.
[10] See the results given in table 12.6 on page 159.

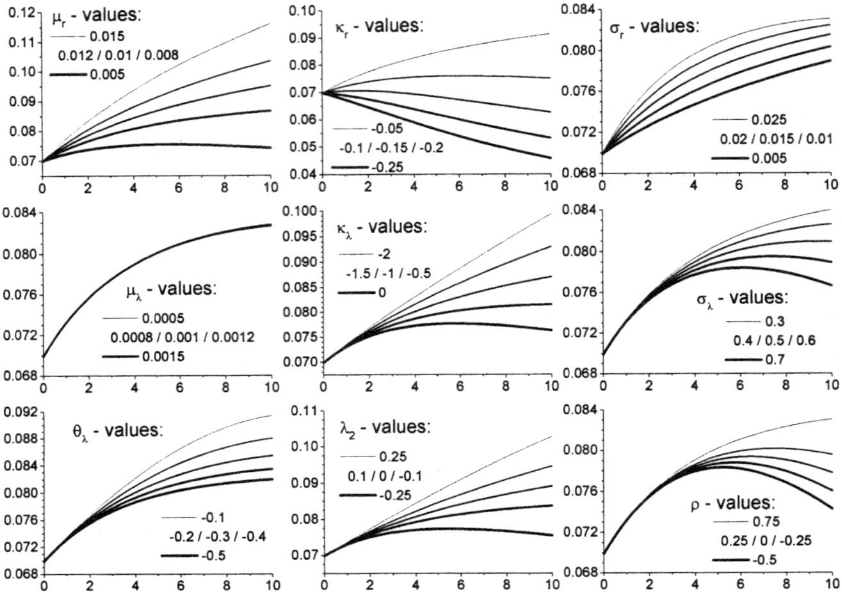

Fig. 10.2. Analysis of the Term Structure of Interest Rates

are all upward sloping.[11] These values for the state variables correspond to the scenario 'risk averse' and '$r < r_l$' reported in the eighth column of table 10.1. In the table we present the effects of positive changes in the parameter values for all possible scenarios of the state variables. When a rise in parameter values has an increasing effect on the yield curve it is denoted by '$\Uparrow$'; decreasing and neglectable effects are symbolized by '$\Downarrow$' and '$\approx$', respectively.

In explaining the contents of table 10.1 we begin with the influences of the parameters μ_r, μ_λ, and λ_2. We consider these parameters as level parameters since they show up as additive constants in the drift terms of equations (10.27) and (10.28). They all have a constant influence on the yield curve throughout the different scenarios. The parameter μ_r gives the short rate higher presumed values in the future in thus it has an increasing effect on the yield curve. The additive terms in the process of the market price of risk are μ_λ and λ_2. Numerically positive changes in μ_λ result in lower yields, but within the empirically estimated values for the parameter we only observe neglectable effects. The opposite influence is attained by λ_2 since it enters the stochastic process with an opposite sign. The values for these parameters both make the expected value of the market price of risk either rise or fall. This further

[11] The long-run means of the state variables in the data set are $\mathbf{x}_l = [0.0950, -0.1311]'$ which are based on the US interest data; for further analysis of the state variables see section 12.3.3.

Table 10.1. Effects of Increasing Parameter Values on the Yield Curve

<table>
<tr><td rowspan="4">Model Para-meters ψ</td><td colspan="9">Market Price of Risk</td></tr>
<tr><td colspan="3">Risk Taking ($\lambda > 0$)</td><td colspan="3">Risk Neutral ($\lambda = 0$)</td><td colspan="3">Risk Averse ($\lambda < 0$)</td></tr>
<tr><td colspan="9">Short Interest Rate</td></tr>
<tr><td>$r < r_l$</td><td>$r = r_l$</td><td>$r > r_l$</td><td>$r < r_l$</td><td>$r = r_l$</td><td>$r > r_l$</td><td>$r < r_l$</td><td>$r = r_l$</td><td>$r > r_l$</td></tr>
<tr><td>μ_r</td><td colspan="9">$\upuparrows$</td></tr>
<tr><td>κ_r</td><td>$\downdownarrows$</td><td>$\upuparrows$</td><td>$\upuparrows$</td><td colspan="3">$\upuparrows$</td><td colspan="3">$\upuparrows$</td></tr>
<tr><td>σ_r</td><td colspan="3">$\downdownarrows$</td><td>$\downdownarrows$</td><td>$\approx$</td><td>$\upuparrows$</td><td colspan="3">$\upuparrows$</td></tr>
<tr><td>μ_λ</td><td colspan="9">$\approx$</td></tr>
<tr><td>κ_λ</td><td>$\upuparrows$</td><td>$\downdownarrows$</td><td>$\downdownarrows$</td><td colspan="3">$\downdownarrows$</td><td colspan="3">$\downdownarrows$</td></tr>
<tr><td>σ_λ</td><td colspan="9">$\downdownarrows$</td></tr>
<tr><td>θ_λ</td><td colspan="3">$\downdownarrows$</td><td>$\downdownarrows$</td><td>$\upuparrows\downdownarrows$</td><td>$\upuparrows$</td><td colspan="3">$\upuparrows$</td></tr>
<tr><td>λ_2</td><td colspan="9">$\upuparrows$</td></tr>
<tr><td>ρ</td><td colspan="9">$\upuparrows$</td></tr>
<tr><td colspan="10">Effects on Yield Curve: $\upuparrows$: increasing $\downdownarrows$: decreasing $\approx$: neglectable</td></tr>
</table>

gives a feedback on the short rate which is lower under the measure $\mathbb{Q}$. Thus, higher values for the market price of risk lower the term structure of interest rates as we demonstrated in the previous section.

The next parameters we analyze are the direct factor loadings κ_r, κ_λ, and θ_λ in the stochastic processes of equations (10.27) and (10.28). With respect to the short rate higher values of κ_r assume higher expectations of the investors in the future in that it has an increasing effect on the yield curve. Only in the case of high values of λ we observe a contrary reaction. The market price of risk is directly influenced by the parameters κ_λ and θ_λ. With positive changes in κ_λ we generally lower the expectations in the market price of risk which makes the investors require higher returns in the future. This results in an increase of the yield curve. The impact of the level of the market price on itself is modeled by θ_λ. Its influence on the yield curve varies in dependence of the sign of λ_t, i.e. the current risk attitude of investors. In risk taking scenarios we attain a negative influence and with risk averse investors we obtain rising yield curves. For risk-neutral investors with the current short rate being at its long-run level the short rates are pushed up whereas the long rates are lowered.

Moreover, the state variables are modeled with a possible correlation factor ρ. With high values for this parameter it is more likely to observe high values of the short rate whenever the risk attitude is high. It only influences the term

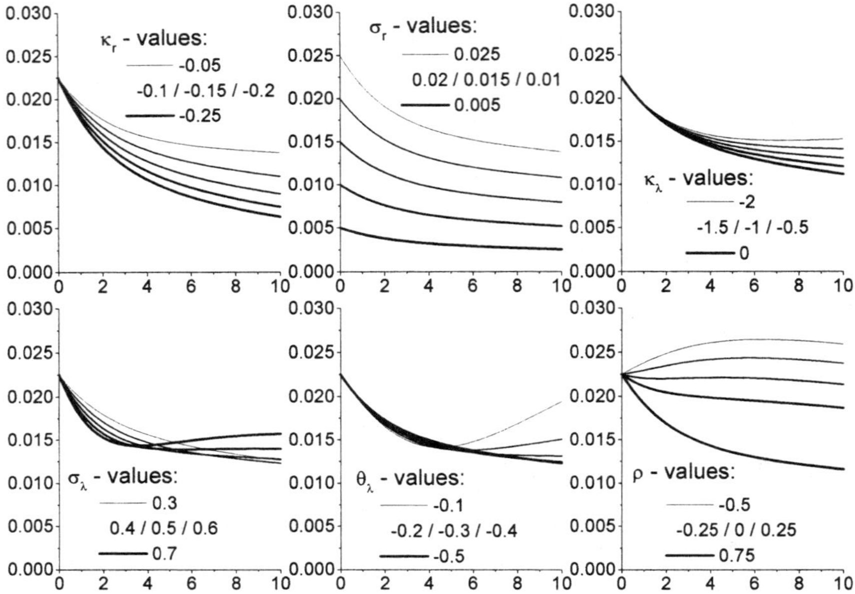

Fig. 10.3. Typical Patterns of the Term Structure of Volatilities

structure of interest rates by the term $C(t,T)$ according to equation (10.18). Through all different scenarios of the state variables we observe increasing yields with positive changes in the correlation coefficient.

Finally, we examine the influences of the volatility parameters σ_r and σ_λ. The volatility of the short rate does not directly impact the expected value of the short rate. However, with higher values for σ_r risk averse investors need to be compensated with higher returns. Thus, the yield curve increases with higher σ_r-values. Risk taking investors on the other side love to have more volatility in that they are satisfied with lower returns. The volatility σ_λ on the risk premium models the frequency of changing risk attitudes of the investors through time. Thereby, high fluctuations result in lower levels of the yield curve.

Now, we move on to consider the comparative analysis of the term structure of *spot rate volatilities*. The possible cases given by the standard parameter values for the possible cases of volatility structures are shown in figure 10.3. From equation (10.21) we see that the exogenous parameters that influence the volatility structure are given by the reduced number of $\psi = \{\kappa_r, \sigma_r, \kappa_\lambda, \sigma_\lambda, \theta_\lambda, \rho\}$. Thus, aside from the state variables the level parameters μ_r, μ_λ, and λ_2 do not impact the volatility structure as it is expected from finance theory.

The volatility structures show us how the short rate volatility σ_r is transformed and extrapolated towards yields of higher maturities. At the ordinate we coincide with the value of the short rate volatility $\sigma_r = 0.0225$. This fact can conveniently be seen from the σ_r−graph in figure 10.3 as the levels of the volatility curves change with different values for σ_r. Herein, the volatilities for the ten year rate range between 51 to 55 percent of the short rate value, i.e. the relative volatility is about 50 percent.

All other parameters examined show a positive effect on the yield curve. With positive changes in the parameter values the volatilities are shifted upwards for longer maturities. In the fourth graph we observe that we are able to let the relative volatility decrease faster over the spectrum of maturities by altering the values for the volatility σ_λ of the market price of risk. The last two graphs show volatility structures that increase with higher time to maturities which is a rather unrealistic phenomenon when compared to empirical findings. However, these patterns are obtained with altering the values for the parameters θ_λ and ρ and at the same time holding the other parameters constant at their standard values. This is only done for explaining comparative results and is not observable in the empirical results.[12]

[12] Compare the results presented on the empirical estimations in chapter 12.

11

Risk Management and Derivatives Pricing

11.1 Management of Interest Rate Risk

After examining the implications of our theoretical term structure model on the different term structures that are relevant in pricing term structure derivatives we now answer the question on how the model can be used in risk management. The management of interest rate risk is especially concerned with rebalancing a fixed-income portfolio exposed to interest rate risk due to the desired risk return characteristics. The types of interest rate risk considered to be of relevance in interest rate risk management are generally considered to be: (i) *Market risk* is the risk of changing prices due to general changes of the overall level of interest rates on default free securities, (ii) *yield curve risk* is considered the risk associated with non-parallel shifts in the yield curve, i.e. a reshaping of the yield curve due to, for example, steepening, flattening, or twisting, and (iii) *credit risk* which is related to altering security prices caused by changes in the creditworthiness of the issuer.

The oldest risk, risk managers have aimed at, is the market risk for which the conventional concepts of duration and convexity enabled them to successfully immunize their portfolios against market risk.[1] The yield curve risk asked for better measures of interest rate risk beyond capturing infinitesimal parallel shifts in the yield curve. Here, especially, the key rate durations of Ho (1992) gained popularity.[2] However, to cope with this risk type managers already use interest rate derivative securities such as swap contracts, or cap and floor

[1] See, for example, Macaulay (1938) and Bierwag, Kaufman, and Toevs (1983).

[2] Other recent concepts include the functional duration from Klaffky, Ma, and Nozari (1992) and the partial duration by Waldman (1992).

agreements. Lastly, the credit risk is the newest risk exposure fixed-income managers start to deal with.[3]

The risks captured by our proposed term structure model include the market and the yield curve risk whereas it is not capable to incorporate any risk associated with default risky securities. As we analyzed in section 10.4 the first model factor can be seen as level factor and thus attributed to describe general market movements of interest rates. With the market price of risk factor we additionally consider the steepness of the term structure of interest rates and thereby capture the yield curve risk. The two factors can successfully be applied to the concept of duration using factor durations to manage the market and yield curve risk.

Definition 11.1 (Factor Duration) [4] *Given the price of a coupon bond* $P^c(t,T)$ *the factor duration is defined by*

$$\begin{aligned} D_F &= -\frac{1}{P^c(t,T)} \frac{\partial P^c(t,T)}{\partial F} \\ &= -\frac{1}{P^c(t,T)} \sum_{u \in (t;T]} c(u) \frac{\partial P(t,u)}{\partial F} \end{aligned}$$

for a factor F.

With respect to the two factors in our term structure model we get the factor durations

$$D_r = -\frac{1}{P^c(t,T,\mathbf{x})} \sum_{u \in (t;T]} c(u) \frac{\partial P^c(t,u,\mathbf{x})}{\partial r}, \text{ and} \tag{11.1}$$

$$D_\lambda = -\frac{1}{P^c(t,T,\mathbf{x})} \sum_{u \in (t;T]} c(u) \frac{\partial P^c(t,u,\mathbf{x})}{\partial \lambda} \tag{11.2}$$

for the short rate r and the market price of risk λ. The factor durations measure the sensitivity of bond prices to the factors. In the case of a discount bond with $c(T) = 1$ we get the measures of risk exposure as $D_r = A(t,T)$, and $D_\lambda = B(t,T)$, where the functions $A(t,T)$ and $B(t,T)$ are from equations (10.12) and (10.13).

They duration measures of equations (11.1) and (11.2) can directly be derived from the stochastic dynamics of coupon bonds. Starting with the infinitesimal time changes of a discount bond $P(t,T,\mathbf{x}) = \exp[-y(t,T)(T-t)]$

[3] For a comparative analysis of current credit risk models used as internal models by banks see, for example, Crouhy, Galai, and Mark (2000).

[4] See, for example, Wu (2000) and Chen (1996a).

based on equation (9.1) using Itô's formula yields:

$$\frac{dP(t,T,\mathbf{x})}{P(t,T,\mathbf{x})} = \mu_P dt + \boldsymbol{\eta} d\mathbf{W}_t^{\mathbb{P}}, \tag{11.3}$$

$$\text{with } \mu_P = \frac{1}{P(t,T,\mathbf{x})}\left[\frac{1}{2}\sigma_r^2 \frac{\partial^2 P(t,T,\mathbf{x})}{\partial r^2} + \rho\sigma_r\sigma_\lambda \frac{\partial^2 P(t,T,\mathbf{x})}{\partial r \partial \lambda}\right.$$
$$+\frac{1}{2}\sigma_\lambda^2 \frac{\partial^2 P(t,T,\mathbf{x})}{\partial \lambda^2} + (\mu_r + \kappa_r r)\frac{\partial P(t,T,\mathbf{x})}{\partial r}$$
$$\left. + (\mu_\lambda + \kappa_\lambda r)\frac{\partial P(t,T,\mathbf{x})}{\partial \lambda} + \frac{\partial P(t,T,\mathbf{x})}{\partial t}\right], \text{ and}$$
$$\boldsymbol{\eta} = \frac{1}{P(t,T,\mathbf{x})}\left[\sigma_r \frac{\partial P(t,T,\mathbf{x})}{\partial r} \quad \sigma_\lambda \frac{\partial P(t,T,\mathbf{x})}{\partial \lambda}\right]'.$$

To obtain the dynamics of coupon bonds we use equation (11.3):

$$\frac{dP^c(t,T,\mathbf{x})}{P^c(t,T,\mathbf{x})} = \sum_{u\in(t;T]} c(u)\mu_P dt + D_r\sigma_r dW_{r,t}^{\mathbb{P}} + D_\lambda\sigma_\lambda dW_{\lambda,t}^{\mathbb{P}}, \tag{11.4}$$

where we already substituted for the duration measures of equations (11.1) and (11.2). From equation (11.4) we see that investors who want to immunize their portfolios against market and yield curve risk will construct a portfolio that has the same factor durations as the relevant fixed-income benchmark.

In the next section we show how to price selected derivative contracts relevant in risk management using the EMM approach: Options on discount bonds, interest rate swap contracts, and interest rate cap and floor agreements.

11.2 Pricing Interest Rate Derivatives

With the value of a pure discount bond we have priced the first interest rate security in section 10.1. Based on this price we have further derived the various term structures and correlation interdependencies of the yields that were relevant in order to characterize the term structure model. Now, we want to examine the prices of further specifications of derivatives on the term structure. Therefore, we use the method of *martingale pricing* (EMM approach) which will be shown to be a powerful tool in pricing interest rate derivative securities.[5]

[5] For the EMM approach see section 1.2.2 and for a broad variety of applications of this method see, for example, Musiela and Rutkowski (1997).

In the case of discount bonds we obtain an explicit closed-form solution to the fundamental partial differential equation of equation (9.4); the specific boundary value was set to $P(T, T, \mathbf{x}) = 1$. To price other contingent claims we appropriately need to change the boundary condition. The different solution to the differential equation then reflects the specified contingent claim.[6] However, instead of deriving the value for the contingent claim from the partial differential equation we can equivalently obtain its value by calculating the expected discounted pay-off value. For example, the previously derived discount bond formula is the solution to $\mathbb{E}^{\mathbb{Q}}\left[\exp - \int_t^T r(\zeta)\, d\zeta \middle| \mathcal{F}_t\right]$. The peculiarity about the calculation of such expectations is that it is done not with respect to the original measure $\mathbb{P}$ but with respect to the equivalent martingale measure $\mathbb{Q}$. The relevant martingale measure is found by applying the right numéraire asset which makes the contingent claim a martingale.[7] In evaluating the prices of equity options one conveniently works with the money market account $B(t, T)$ as numéraire. However, in pricing term structure derivatives we use the *forward martingale measure*, which we thoroughly introduce, with the EMM approach in the coming sections

11.2.1 Bond Options

In this section we examine the most basic of all derivatives on discount bonds, namely call and put options. After defining the specific contract of a European call option, we derive its todays fair value. Finally, we transfer the pricing result of call options to value put options which can be easily obtained using a theoretical parity relation between call and put prices.

Definition 11.1 (European Call Option) *A European style call option with expiration date T on a zero-coupon bond which matures at time $U \geq T$ is defined by the contingent payoff*

$$\Pi(T, T, \mathbf{x}) = (P(T, U) - K)^+ ,$$

which is the value of the call option C_T at time T. It can only be exercised at the expiration date T.[8]

[6] A method studied extensively for different contingent claims by, for example, Wilmott, Dewynne, and Howison (1993).

[7] This results in discounted asset prices being martingales with respect to $\mathbb{Q}$; see Harrison and Pliska (1981) and Geman, Karoui, and Rochet (1995).

[8] From now onwards we drop the notation for the $\mathbf{x}$-dependence of discount bonds, i.e. we denote their price by $P(t, T)$.

From this definition we know the terminal condition of the derivative contract which we use in conjunction with the general value of a derivative contract as given in equation (1.8). In order to get a value of the call at a date prior to the expiration T of the option, we need to calculate the discounted expected value of the option under the equivalent martingale measure $\mathbb{Q}$, i.e.

$$C_t = \mathbb{E}^{\mathbb{Q}}\left[e^{-\int_t^T r(\zeta)d\zeta} \left(P(T,U) - K\right)^+ \middle| \mathcal{F}_t \right]. \tag{11.5}$$

Now, we use the fact that a call option is a piecewise linear instrument, which gives us the possibility of using *indicator functions* from probability theory - denoted by $\mathbf{1}_{\{\dots\}}$ - within our specification in equation (11.5). This yields

$$\begin{aligned} C_t = {} & \mathbb{E}^{\mathbb{Q}}\left[e^{-\int_t^T r(\zeta)d\zeta} P(T,U)\, \mathbf{1}_{\{P(T,U)\geq K\}} \middle| \mathcal{F}_t \right] \\ & - K\mathbb{E}^{\mathbb{Q}}\left[e^{-\int_t^T r(\zeta)d\zeta} \mathbf{1}_{\{P(T,U)\geq K\}} \middle| \mathcal{F}_t \right]. \end{aligned} \tag{11.6}$$

However, we cannot split the terms in the first expectation since they are not independent under $\mathbb{Q}$. To further analyze the expectations, we use the powerful idea of choosing the optimal numéraire in the following when pricing and hedging a given contingent claim as described in Geman, Karoui, and Rochet (1995).[9] They change measure from the accumulator measure in use to a new equivalent *forward measure* by introducing a discount bond as a practical numéraire.

Definition 11.2 (Forward Martingale Measure) *For a forward martingale measure $\mathbb{Q}^T$, equivalent to the measure $\mathbb{Q}$, there exists a random variable ζ^T for any random variable X, with the property*

$$\int_{\Omega} X(\omega)\, d\mathbb{Q}^T(\omega) = \int_{\Omega} X(\omega)\, \zeta^T d\mathbb{Q}(\omega).$$

The function ζ^T, often denoted by $\frac{d\mathbb{Q}^T}{d\mathbb{Q}}$, is the Radon-Nikodým derivative of $\mathbb{Q}^T$ with respect to $\mathbb{Q}$ defined on $\mathcal{F}_t$ for $0 \leq t \leq T$. The Radon-Nikodým

[9] The technique of *changing numéraire* was developed by Geman (1989) for the general case of stochastic interest rates. Jamshidian (1989) implicitly uses this technique in the Gaussian interest rate framework of Vasicek (1977).

derivative $\frac{d\mathbb{Q}^T}{d\mathbb{Q}}$ *is given by the ratio of the numéraires* N *in use, i.e.* $\zeta^T = N^{\mathbb{Q}^T}/N^{\mathbb{Q}}$. *For the old numéraire we use the money market account*

$$N^{\mathbb{Q}} = \frac{B(0,t)}{B(0,0)}$$

with $B(0,0) = 1$, *whereas the numéraire for the measure* $\mathbb{Q}^T$ *is given by the discount bond maturing at date* T

$$N^{\mathbb{Q}^T} = \frac{P(t,T)}{P(0,T)},$$

which is normalized to force a unit value at time $t = 0$.

Conveniently, we now change measure from the risk adjusted measure $\mathbb{Q}$ to the equivalent *forward martingale measures* $\mathbb{Q}^U$ and $\mathbb{Q}^T$, which yield the change of measure or *likelihood ratio processes*

$$\zeta_t^U = \mathbb{E}^{\mathbb{Q}}\left[\left.\frac{d\mathbb{Q}^U}{d\mathbb{Q}}\right| \mathcal{F}_t\right] = \frac{P(t,U)}{B(0,t)\,P(0,U)}, \text{ and}$$
$$\zeta_t^T = \mathbb{E}^{\mathbb{Q}}\left[\left.\frac{d\mathbb{Q}^T}{d\mathbb{Q}}\right| \mathcal{F}_t\right] = \frac{P(t,T)}{B(0,t)\,P(0,T)};$$

these stochastic processes are martingales with respect to the measure $\mathbb{Q}$.[10]

For the further calculations we make use of the following abstract version of the Bayes's formula, the 'Bayes's formula for change of numéraire'.

Theorem 11.3 (Bayes's Formula). *For a process* Y *adapted to* $\mathcal{F}_t$ *and the Radon-Nikodým derivative* $d\mathbb{Q}_1/d\mathbb{Q}_2 = \zeta$ *we have*

$$\mathbb{E}^{\mathbb{Q}_1}[Y|\mathcal{F}_t] = \frac{\mathbb{E}^{\mathbb{Q}_2}[Y\zeta|\mathcal{F}_t]}{\mathbb{E}^{\mathbb{Q}_2}[\zeta|\mathcal{F}_t]} = \frac{1}{\zeta_t}\mathbb{E}^{\mathbb{Q}_2}[Y\zeta|\mathcal{F}_t]$$

or equivalently

$$\mathbb{E}^{\mathbb{Q}_2}[Y|\mathcal{F}_t] = \frac{\mathbb{E}^{\mathbb{Q}_1}\left[Y\zeta^{-1}|\mathcal{F}_t\right]}{\mathbb{E}^{\mathbb{Q}_1}\left[\zeta^{-1}|\mathcal{F}_t\right]} = \zeta_t\mathbb{E}^{\mathbb{Q}_1}\left[Y\zeta^{-1}|\mathcal{F}_t\right]$$

under the usual conditions.[11]

---

[10] For likelihood ratio processes in interest rate theory see, for example, Björk (1997).

[11] For applying Bayes's formula to financial pricing issues see, for example, Dothan (1990, p. 288).

Coming back to equation (11.6) we calculate the two expectations separately. For the first expectation we use the Radon-Nikodým derivatives ζ_t^U and ζ_T^U to change measure from $\mathbb{Q}$ to the U-forward measure $\mathbb{Q}^U$:

$$\begin{aligned}
\mathbb{E}_1^{\mathbb{Q}} &= \zeta_t^U \mathbb{E}^{\mathbb{Q}^U}\left[e^{-\int_t^T r(\zeta)d\zeta} P(T,U)\, \zeta_T^{U-1} \mathbf{1}_{\{P(T,U)\geq K\}} \middle| \mathcal{F}_t \right] \\
&= \frac{P(t,U)}{B(0,t)\,P(0,U)} \mathbb{E}^{\mathbb{Q}^U}\left[e^{-\int_t^T r(\zeta)d\zeta} P(T,U) \right. \\
&\quad \left. \frac{B(0,T)\,P(0,U)}{P(T,U)} \mathbf{1}_{\{P(T,U)\geq K\}} \middle| \mathcal{F}_t \right] \\
&= \frac{P(t,U)\,P(0,U)}{P(0,U)} \mathbb{E}^{\mathbb{Q}^U}\left[e^{-\int_t^T r(\zeta)d\zeta} \frac{B(0,T)}{B(0,t)} \mathbf{1}_{\{P(T,U)\geq K\}} \middle| \mathcal{F}_t \right] \\
&= P(t,U)\, \mathbb{E}^{\mathbb{Q}^U}\left[\mathbf{1}_{\{P(T,U)\geq K\}} \middle| \mathcal{F}_t \right].
\end{aligned}$$

Similarly, to make a change from the accumulator measure to the T-forward measure we use ζ_t^T and ζ_T^T:

$$\begin{aligned}
\mathbb{E}_2^{\mathbb{Q}} &= \zeta_t^T \mathbb{E}^{\mathbb{Q}^T}\left[e^{-\int_t^T r(\zeta)d\zeta} \zeta_T^{T-1} \mathbf{1}_{\{P(T,U)\geq K\}} \middle| \mathcal{F}_t \right] \\
&= \frac{P(t,T)}{B(0,t)\,P(0,T)} \mathbb{E}^{\mathbb{Q}^T}\left[e^{-\int_t^T r(\zeta)d\zeta} \frac{B(0,T)\,P(0,T)}{1} \mathbf{1}_{\{P(T,U)\geq K\}} \middle| \mathcal{F}_t \right] \\
&= \frac{P(t,T)\,P(0,T)}{P(0,T)} \mathbb{E}^{\mathbb{Q}^T}\left[e^{-\int_t^T r(\zeta)d\zeta} \frac{B(0,T)}{B(0,t)} \mathbf{1}_{\{P(T,U)\geq K\}} \middle| \mathcal{F}_t \right] \\
&= P(t,T)\, \mathbb{E}^{\mathbb{Q}^T}\left[\mathbf{1}_{\{P(T,U)\geq K\}} \middle| \mathcal{F}_t \right].
\end{aligned}$$

When we insert the results for the expectations $\mathbb{E}_1^{\mathbb{Q}}$ and $\mathbb{E}_2^{\mathbb{Q}}$ back into equation (11.6), we get

$$\begin{aligned}
C_t &= P(t,U)\, \mathbb{E}^{\mathbb{Q}^U}\left[\mathbf{1}_{\{P(T,U)\geq K\}} \middle| \mathcal{F}_t \right] - P(t,T)\, K \mathbb{E}^{\mathbb{Q}^T}\left[\mathbf{1}_{\{P(T,U)\geq K\}} \middle| \mathcal{F}_t \right] \\
&= P(t,U)\, \mathbb{Q}^U \left(P(T,U) \geq K\right) - P(t,T)\, K \mathbb{Q}^T \left(P(T,U) \geq K\right). \qquad (11.7)
\end{aligned}$$

For the purpose of further calculating the probabilities in equation (11.7) we use relative discount bond prices. For the calculation of the *first probability*, which is defined in the U-forward measure, we work in terms of the $P(t,U)$ bond as numéraire, defining us the relative price $Y_t^U = P(t,T) \,/\, P(t,U)$:

$$\mathbb{Q}^U\left(P(T,U) \geq K\right) = \mathbb{Q}^U\left(\frac{P(T,T)}{P(T,U)} \leq \frac{1}{K}\right) = \mathbb{Q}^U\left(Y_T^U \leq \frac{1}{K}\right). \qquad (11.8)$$

Using the explicit formula for discount bonds in equation (10.1) we calculate the forward bond price Y_t^T as

$$Y_t^T = e^{[A(t,U)-A(t,T)]r+[B(t,U)-B(t,T)]\lambda+C(t,U)-C(t,T)}.$$

Applying Itô's formula based on the $r(t)-$ and $\lambda(t)-$dynamics of equation (9.2) we get the dynamics of Y_t^T under the measure $\mathbb{Q}$:

$$\begin{aligned} dY_t^T &= \frac{\partial Y_t^T}{\partial t}dt + \frac{\partial Y_t^T}{\partial r}dr + \frac{\partial Y_t^T}{\partial \lambda}d\lambda \\ &\quad + \frac{1}{2}\frac{\partial^2 Y_t^T}{\partial r^2}dr^2 + \frac{\partial^2 Y_t^T}{\partial r \partial \lambda}dr d\lambda + \frac{1}{2}\frac{\partial^2 Y_t^T}{\partial \lambda^2}d\lambda^2 \\ \frac{dY_t^T}{Y_t^T} &= (\ldots)\,dt + \sigma_r\left(A(t,U) - A(t,T)\right)dW_r^{\mathbb{Q}} \\ &\quad + \sigma_\lambda\left(B(t,U) - B(t,T)\right)dW_\lambda^{\mathbb{Q}}. \end{aligned}$$

Then, under the equivalent U-forward measure Y_t^T is a martingale which can be described by

$$\begin{aligned} dY_t^T &= \sigma_{Y,r}Y_t^T dW_r^{\mathbb{Q}^U} + \sigma_{Y,\lambda}Y_t^T dW_\lambda^{\mathbb{Q}^U}, \qquad (11.9) \\ \text{with } \sigma_{Y,r} &= \sigma_r\left(A(t,U) - A(t,T)\right), \text{ and} \\ \sigma_{Y,\lambda} &= \sigma_\lambda\left(B(t,U) - B(t,T)\right). \end{aligned}$$

From equation (11.9) we see that Y_t^T follows a log-normal distribution. Applying Itô's formula to the function $\ln Y_t^T$ - with partial derivatives $\partial \ln Y_t^T / \partial Y_t^T = 1/Y_t^T$ and $\partial^2 \ln Y_t^T / \partial^2 Y_t^T = -\,1/{Y_t^T}^2$ - we obtain the $\mathbb{Q}^U$-dynamics as

$$\begin{aligned} d\ln Y_t^T &= \frac{\partial \ln Y_t^T}{\partial Y_t^T}dY_t^T + \frac{1}{2}\frac{\partial^2 \ln Y_t^T}{\partial^2 Y_t^T}\left(dY_t^T\right)^2 \\ &= \sigma_{Y,r}dW_r^{\mathbb{Q}^U} + \sigma_{Y,\lambda}dW_\lambda^{\mathbb{Q}^U} \\ &\quad -\frac{1}{2}\left(\sigma_{Y,r}^2 + 2\rho\sigma_{Y,r}\sigma_{Y,\lambda} + \sigma_{Y,\lambda}^2\right)dt. \qquad (11.10) \end{aligned}$$

We now integrate equation (11.10) over the time period $[t,T]$:

$$\ln Y_T^T = \ln Y_t^T + \int_t^T \sigma_{Y,r} dW_r^{\mathbb{Q}^U} + \int_t^T \sigma_{Y,\lambda} dW_\lambda^{\mathbb{Q}^U}$$
$$-\frac{1}{2}\int_t^T \sigma_{Y,r}^2 + 2\rho\sigma_{Y,r}\sigma_{Y,\lambda} + \sigma_{Y,\lambda}^2 d\zeta.$$

Therein, the stochastic integrals are Gaussian random variables with zero mean. Since Itô integrals have an expectation of zero, the sum of the two random variables possesses a zero mean. The variance of the two integrals can be calculated with Itô's isometry.

Theorem 11.4 (The Itô Isometry). [12] *For a function $f(t,\omega)$ we have*

$$\mathbb{E}\left[\left(\int_t^T f(t,\omega)\, dW\right)^2 \middle| \mathcal{F}_t\right] = \mathbb{E}\left[\int_t^T f^2(t,\omega)\, dt \middle| \mathcal{F}_t\right].$$

Using this theorem we can calculate the variance as:

$$\begin{aligned}
\upsilon_Y^2 &= \mathbb{E}^{\mathbb{Q}^U}\left[\left(\int_t^T \sigma_{Y,r} dW_r^{\mathbb{Q}^U} + \int_t^T \sigma_{Y,\lambda} dW_\lambda^{\mathbb{Q}^U}\right)^2 \middle| \mathcal{F}_t\right] \\
&= \int_t^T \left(\sigma_{Y,r}^2 + 2\rho\sigma_{Y,r}\sigma_{Y,\lambda} + \sigma_{Y,\lambda}^2\right)^2 d\zeta \\
&= \int_t^T \sigma_r^2 \left(A(\zeta,U) - A(\zeta,T)\right)^2 \\
&\quad +2\rho\sigma_r\sigma_\lambda \left(A(\zeta,U) - A(\zeta,T)\right)\left(B(\zeta,U) - B(\zeta,T)\right) \\
&\quad +\sigma_\lambda^2 \left(B(\zeta,U) - B(\zeta,T)\right) d\zeta. \qquad (11.11)
\end{aligned}$$

The expression for υ_Y^2 gained in equation (11.11) actually is the variance of the relative bond price Y_t^U. Thereby, we gained all relevant information to calculate the probability from equation (11.8):

[12] See, for example, Oksendal (2000).

$$\begin{aligned}
\mathbb{Q}^U\left(Y_T^U \leq \frac{1}{K}\right) &= \mathbb{Q}^U\left(\ln Y_T^U \leq \ln \frac{1}{K}\right) \\
&= \mathbb{Q}^U\left(\ln Y_t^U - \int_t^T \sigma_{Y,r} dW_r^{\mathbb{Q}^U} - \int_t^T \sigma_{Y,\lambda} dW_\lambda^{\mathbb{Q}^U}\right. \\
&\quad \left. - \frac{1}{2}\int_t^T \sigma_Y^2 d\zeta \leq \ln \frac{1}{K}\right) \\
&= \mathbb{Q}^U\left(\int_t^T \sigma_{Y,r} dW_r^{\mathbb{Q}^U} + \int_t^T \sigma_{Y,\lambda} dW_\lambda^{\mathbb{Q}^U}\right. \\
&\quad \left. \leq \ln \frac{1}{KY_t^U} + \frac{1}{2}\upsilon_Y^2\right).
\end{aligned}$$

Using the fact that the Itô integrals in this equation is Gaussian with $N\left(0, \upsilon_Y^2\right)$, we finally get

$$\mathbb{Q}^T\left(Y_T^U \leq \frac{1}{K}\right) = N\left(\frac{\ln \frac{P(t,U)}{KP(t,T)} + \frac{1}{2}\upsilon_Y^2}{\upsilon_Y}\right).$$

Using similar arguments we now calculate the *second probability* in equation (11.7). Here, we naturally work in terms of relative prices with respect to the discount bond $P(t,T)$ because of the T-forward measure $\mathbb{Q}^T$. We use the forward price defined by $Z_t^T = P(t,U)/P(t,T)$:

$$\mathbb{Q}^T\left(P(T,U) \geq K\right) = \mathbb{Q}^T\left(\frac{P(T,U)}{P(T,T)} \geq K\right) = \mathbb{Q}^T\left(Z_T^T \geq K\right). \quad (11.12)$$

For the stochastic dynamics of Z_t^U we get the analogous expression to equation (11.9) by using Itô's formula which yields

$$\begin{aligned}
dZ_t^T &= \sigma_{Z,r} Z_t^T dW_r^{\mathbb{Q}^U} + \sigma_{Z,\lambda} Z_t^T dW_\lambda^{\mathbb{Q}^U}, \quad (11.13) \\
\text{with } \sigma_{Z,r} &= \sigma_r\left(A(t,T) - A(t,U)\right), \text{ and} \\
\sigma_{Z,\lambda} &= \sigma_\lambda\left(B(t,T) - B(t,U)\right).
\end{aligned}$$

Comparing the result of equation (11.13) with equation (11.9), we note the equivalence of the volatility terms just by a reversion of the sign, i.e. $\sigma_{Z,r} = -\sigma_{Y,\lambda}$ and $\sigma_{Z,r} = -\sigma_{Y,\lambda}$. Further, we obtain

$$\ln Z_T^T = \ln Z_t^T + \int_t^T \sigma_{Z,r} dW_r^{\mathbb{Q}^U} + \int_t^T \sigma_{Z,\lambda} dW_\lambda^{\mathbb{Q}^U}$$
$$-\frac{1}{2}\int_t^T \sigma_{Z,r}^2 + 2\rho\sigma_{Z,r}\sigma_{Z,\lambda} + \sigma_{Z,\lambda}^2 d\zeta.$$

Thereby, we result in the same variance for the forward bond price $P(T,U)$ as in the previous case of the relative price Y_t^U, i.e. we get the identity relationship

$$\upsilon^2 = \upsilon_Z^2 = \upsilon_Y^2.$$

With this result, we obtain for the probability in equation (11.12)

$$\begin{aligned}\mathbb{Q}^T\left(Z_T^T \geq K\right) &= \mathbb{Q}^T\left(\ln Z_T^T \geq \ln K\right)\\ &= \mathbb{Q}^T\left(\ln Z_t^T + \int_t^T \sigma_{Z,r} dW_r^{\mathbb{Q}^U} + \int_t^T \sigma_{Z,\lambda} dW_\lambda^{\mathbb{Q}^U}\right.\\ &\quad\left. -\frac{1}{2}\int_t^T \sigma_Z^2 d\zeta \geq \ln K\right)\\ &= \mathbb{Q}^T\left(\ln\frac{Z_t^T}{K} - \frac{1}{2}\upsilon^2 \geq \int_t^T \sigma_{Y,r} dW_r^{\mathbb{Q}^U} + \int_t^T \sigma_{Y,\lambda} dW_\lambda^{\mathbb{Q}^U}\right)\\ &= N\left(\frac{\ln\frac{P(t,U)}{KP(t,T)} - \frac{1}{2}\upsilon^2}{\upsilon}\right).\end{aligned}$$

Restating our results, we have shown that the value of a *European call option* can be calculated within our term structure framework using the closed-form expression

$$C_t = P(t,U)\,N(d_1) - P(t,T)\,KN(d_2)\,,\text{ where} \tag{11.14}$$
$$d_{1,2} = \frac{\ln\frac{P(t,U)}{KP(t,T)} \pm \frac{1}{2}\upsilon^2}{\upsilon} \text{ and } \upsilon^2 = \int_t^T \left(\sigma_{P(\zeta,T)} - \sigma_{P(\zeta,U)}\right)^2 d\zeta.$$

In order to derive the arbitrage price of the European Put option belonging to the same option series, we conveniently use the put-call parity relationship.

Corollary 11.5 (Put-Call Parity) *The put-call parity relationship for European options C_t and P_t on discount bonds is immediately obtained as*

$$\begin{aligned}
P_t &= \mathbb{E}^{\mathbb{Q}}\left[e^{-\int_t^T r(\zeta)d\zeta} \left(K - P(T,U)\right)^+ \middle| \mathcal{F}_t \right] \\
&= \mathbb{E}^{\mathbb{Q}}\left[e^{-\int_t^T r(\zeta)d\zeta} \left(P(T,U) - K\right)^+ + K - P(T,U) \middle| \mathcal{F}_t \right] \\
&= C_t + KP(t,T) - P(t,U),
\end{aligned}$$

using the equality $(K - P(T,U))^+ = (P(T,U) - K)^+ + K - P(T,U)$.[13]

Thus, we have the price of a *European put option* given by

$$\begin{aligned}
P_t &= P(t,U)\left(N(d_1) - 1\right) - P(t,T)K\left(1 - N(d_2)\right) \\
&= P(t,T)KN(-d_2) - P(t,U)N(-d_1), \text{ where} \qquad (11.15)
\end{aligned}$$

$$d_{1,2} = \frac{\ln \frac{P(t,U)}{KP(t,T)} \pm \frac{1}{2}\upsilon^2}{\upsilon} \text{ and } \upsilon^2 = \int_t^T \left(\sigma_{P(\zeta,T)} - \sigma_{P(\zeta,U)}\right)^2 d\zeta.$$

The pricing formulas of equations (11.14) and (11.15) based on our two-factor term structure model are actually in line with the Gaussian option pricing formula of Geman, Karoui, and Rochet (1995, p. 456). Having gone through the calculations for the two fair option prices we need to remark the advantage of having the price of discount bonds available in closed-form expressions. This especially eases the work with changing numéraires.

11.2.2 Swap Contracts

Besides the prices of put and call options one of the most important interest rate derivative securities are interest rate swaps. In financial history, no other market than the swap market has grown as rapidly since its initial inception in 1981.[14] Until today the interest rate swap markets have also grown in depth in that they offer a wide variety of swap contracts designed to serve specific needs. The variants of swap contracts include swaps with contingency features, such as forward swaps, swap rate locks, and swaptions, swaps that use indices in their contracts, such as basis swaps, and yield curve swaps, and finally swaps with varying notional principals, like amortizing and accreting swaps.[15]

[13] The put-call parity was first described by Stoll (1969).

[14] For the origin and the development of the swap markets see, for example, Das (2004, ch. 1).

[15] For variant forms of interest rate swaps see, for example, Marshall and Kapner (1993, ch. 3).

However, for our purposes we specialize in so called fixed-for-floating interest rate swaps which exchange a stream of varying payments for a stream of fixed amount payments. This type of swap contract is formalized in the following definition in order to be exactly analyzable.

Definition 11.6 (Interest Rate Swap) *A fixed-for-floating interest rate swap is an agreement whereby two parties undertake to exchange a fixed set of payments with a spot swap rate k for a floating set of LIBOR payments with a rate $L(0, T_{j-1}, T_j)$ at known dates T_j, $j = 1, \dots, n$. The value of such a swap settled in arrears with a tenor of $\delta = [T_{j-1}, T_j]$ and a notional principal of N is given by*

$$SWP_t = \delta N \sum_{j=1}^{n} \left(L(0, T_{j-1}, T_j) - k\right) P(t, T_j),$$

i.e. the difference of the present values of the floating and the fixed payments.

Since financial markets give quotes on *swap rates*, i.e. the fixed rates at which financial institutions offer their clients interest rate swap contracts of different maturities, we further derive a valuation formula for the fair spot swap rate. In arbitrage-free markets the fair swap value is given by:

$$\delta N \sum_{j=1}^{n} \left(L(0, T_{j-1}, T_j) - k\right) P(t, T_j) \stackrel{!}{=} 0,$$

i.e. a fair swap contract is worth zero today for both counterparties. Substituting the implicit forward rate $L(0, T_{j-1}, T_j)$ from equation (8.3) we obtain

$$\delta N \sum_{j=1}^{n} \left(\frac{P(t, T_{j-1}) - P(t, T_j)}{\delta P(t, T_j)} - k\right) P(t, T_j) = 0$$

$$\sum_{j=1}^{n} P(t, T_{j-1}) - P(t, T_j) - \delta k P(t, T_j) = 0$$

$$\sum_{j=1}^{n} P(t, T_{j-1}) - P(t, T_j) = k\delta \sum_{j=1}^{n} P(t, T_j).$$

The final equation can be solved for the swap rate

$$k = \frac{1 - P(t, T_n)}{\delta \sum_{j=1}^{n} P(t, T_j)} \tag{11.16}$$

of a fair fixed-for-floating interest rate swap settled in arrears with tenor δ. With our interest rate model we are able to calculate the equilibrium swap rate of equation (11.16) in closed-form. This advantage over numerical solutions is offered since we can use the bond price formula derived in section 10.1 to obtain the required bond prices $P(t, T_j)$.

In chapter 15 we actually implement this approach in that we use the observable swap rates from the very liquid US swap markets to obtain information about the underlying term structure of interest rates. Based on the swap rates we implement an empirical model and calibrate the model to the information inherent in the swap rates.

11.2.3 Interest Rate Caps and Floors

A more complex but nonetheless very popular interest rate derivative offered by financial institutions are interest rate caps.[16] Such a contractual agreement is designed to provide an insurance for a long floating rate position against rising above a certain interest rate level known as the cap rate. The seller of the agreement is obliged to make a cash payment to the buyer if a particular interest rate exceeds a mutually prespecified rate at some future date or dates. Similarly, an interest rate floor is an equivalent contract where the specific interest rate must be below a preassigned level to pay cash.

Definition 11.7 (Interest Rate Cap and Floor) *An interest rate cap (floor) with tenor $\delta = [T_{j-1}, T_j]$ is a series of caplets (floorlets) which resemble European call (put) options that are settled in arrears at maturity dates T_j, $j = 1, \ldots, n$, on the LIBOR interest rate $L(T_{j-1}, T_j)$. The payoff characteristic of a so-called caplet (floorlet) is defined, respectively, as*

$$Cpl(T_j, T_j) = \delta N \left(L(T_{j-1}, T_j) - K_C \right)^+, \text{ and}$$
$$Flt(T_j, T_j) = \delta N \left(K_F - L(T_{j-1}, T_j) \right)^+,$$

in the case of an in arrears settlement on a notional principal of N with a cap (floor) rate of $K_C (K_F)$.

To calculate the price of an interest rate cap, we start with valuing a single caplet. Using the expression for the LIBOR rate in terms of discount bonds from equation (8.3), we can restate the payoff of the caplet at maturity T_j as:

[16] For the pricing of default-free interest rate caps, for example, under the direct approach see Briys, Crouhy, and Schöbel (1991) and under the log-normal approach see Miltersen, Sandmann, and Sondermann (1997).

$$\begin{aligned} Cpl\left(T_j, T_j\right) &= \delta N \left(\frac{1 - P\left(T_{j-1}, T_j\right)}{\delta P\left(T_{j-1}, T_j\right)} - K_C\right)^+ \\ &= N \left(\frac{1}{P\left(T_{j-1}, T_j\right)} - 1 - \delta K_C\right)^+ . \end{aligned}$$

Now, we discount the payoff at time T_j to calculate the value of the caplet at time t. We conveniently work with the substitution $K_C^* = 1 + \delta K_C$:

$$\begin{aligned} Cpl\left(t, T_j\right) &= N \mathbb{E}^{\mathbb{Q}} \left[e^{-\int_t^{T_j} r(\zeta) d\zeta} \left(\frac{1}{P\left(T_{j-1}, T_j\right)} - K_C^*\right)^+ \middle| \mathcal{F}_t \right] \\ &= N \mathbb{E}^{\mathbb{Q}} \left[e^{-\int_t^{T_{j-1}} r(\zeta) d\zeta} \left(\frac{1}{P\left(T_{j-1}, T_j\right)} - K_C^*\right)^+ P\left(T_{j-1}, T_j\right) \middle| \mathcal{F}_t \right] \\ &= N \mathbb{E}^{\mathbb{Q}} \left[e^{-\int_t^{T_{j-1}} r(\zeta) d\zeta} \left(1 - K_C^* P\left(T_{j-1}, T_j\right)\right)^+ \middle| \mathcal{F}_t \right] \\ &= K_C^* N \mathbb{E}^{\mathbb{Q}} \left[e^{-\int_t^{T_{j-1}} r(\zeta) d\zeta} \left(\frac{1}{K_C^*} - P\left(T_{j-1}, T_j\right)\right)^+ \middle| \mathcal{F}_t \right] . \end{aligned} \tag{11.17}$$

Reinterpreting the result from equation (11.17) in light of the characteristic payoff for a put option we can state the price of a single caplet in terms of the fair value of European put options as given in equation (11.15)

$$Cpl\left(t, T_j\right) = K_C^* N P_t \left(K = 1/K_C^*, T = T_{j-1}, U = T_j\right);$$

i.e. its value is equivalent to that of K_C^* put options on a discount bond with a strike of $1/K_C^*$, a maturity of T_j, and a delivery date T_{j-1}. Thus, we can conveniently use our results from the previous section on options on discount bonds to obtain the closed-form solution for an *interest rate caplet*

$$Cpl\left(t, T_j\right) = K_C^* N \left[P\left(t, T_{j-1}\right) K N\left(-d_2\right) - P\left(t, T_j\right) N\left(-d_1\right)\right], \text{ where}$$

$$d_{1,2} = \frac{\ln \frac{P(t, T_j) K_C^*}{P(t, T_{j-1})} \pm \frac{1}{2} \upsilon^2}{\upsilon} \quad \text{and } \upsilon^2 = \int_t^T \left(\sigma_{P(\zeta, T_{j-1})} - \sigma_{P(\zeta, T_j)}\right)^2 d\zeta.$$

Finally, the value of a cap which is a bundle of caplets is obtained by the sum of all caplet values

$$Cap\left(t, T\right) = \sum_{j=1}^{n} Cpl\left(t, T_j\right), \tag{11.18}$$

i.e. it is equivalent to that of a portfolio of European put options on discount bonds. The fair value of an interest rate floor agreement can be derived analogously in closed-form.

12

Calibration to Standard Instruments

12.1 Estimation Techniques for Term Structure Models

The empirical finance literature on the econometric analysis of term structure models provides a variety of estimation methods. With the objective of building an empirical model that comes close to the theoretical model properties we basically distinguish two approaches in the literature: The time-series and the cross-section approaches.[1] Both methods are easy to implement, but suffer from the fact that they only use part of the available information of the market for fixed-income instruments in the estimation procedure. A comparison of the two approaches in an estimation of the two well studied models by Vasicek (1977) and Cox, Ingersoll, and Ross (1985b) on data of the Dutch interest rates market is found in, for example, DeMunnik and Schotman (1994). However, there is growing literature following an advanced approach of a *hybrid estimation* in both, the time-series and the cross-section domains simultaneously. These new attempts of integrating the dynamic properties of the term structure models with their cross-sectional implications using interest rate data of different maturities stem from some general problems and shortcomings when estimating term structure models.

The key difficulty in estimating term structure models is the *unobservability of the underlying factors.* To deal with this issue, the model factors are usually either approximated, transformed or treated as parameters for the estimation procedure: (i) The approximation method is often applied in time-series estimation, where, for example, Chan, Karolyi, Longstaff, and Sanders (1992) and Longstaff and Schwartz (1992) use the one month Treasury bill rate as an approximation for the instantaneous spot rate. This approach can induce significant measurement errors, since we face a possible inconsistency

[1] For an overview see, for example, Campbell, Lo, and MacKinlay (1997, ch. 11).

that the proxy variable is not in line with the short end of the estimated term structure.[2] Also, the instantaneous interest rate does not depend on the risk premium while, for example, the one month proxy rate does. Thus, with the approximation approach we need to deal with theoretical inconsistencies in the estimation. Other works that can be subsumed under this approach include Broze, Scaillet, and Zakoian (1995), the non-parametric approach by Ait-Sahalia (1996), and the stochastic volatility model by Andersen and Lund (1997). (*ii*) With the transformation method Chen and Scott (1993), for example, use the thirteen week Treasury bill rate as a perfect substitute for the short rate. With at least one bond price observed without error they are able to derive a system of equations where they utilize the conditional density of the state variables to estimate the parameters of a multi-factor Cox, Ingersoll, and Ross (1985b) model. Other examples following this approach include Pearson and Sun (1994), Duan (1994), and Duffie and Singleton (1997). (*iii*) Finally, we can treat the spot rate as an unobserved parameter among other parameters. This approach is close to the procedure of inverting the yield curve to extract implied estimates of the underlying factors as known from the implied volatility estimates of the Black and Scholes (1973) formula.[3] Examples of estimating the short rate along with the model parameters in a cross-section analysis include the works by Brown and Dybvig (1986), Titman and Torous (1989), and Brown and Schaefer (1994). However, thereby we ignore the dynamic properties of the interest rates, but we may exploit some information about mispricing of bonds which can lead to trading profits.[4]

A second problem arises from the *risk premia* inherent in the observable prices of fixed-income instruments but not in the instantaneous short rate as discussed, for example, in Chen (1995). For the purpose of bond valuation, we need to obtain parameter estimates for the risk adjusted processes. However, the parameters denoting the risk premia cannot be estimated separately from neither a pure time-series approach nor a pure cross-section analysis only. Schöbel (1995a), for example, uses a two step procedure where he first estimates parameters associated with the dynamic behavior of the factors from time-series data and identifies the risk premium in a second step from a cross-section sample of the term structure.

---

[2] For a study on proxies of the short rate see, for example, Chapman, Long, and Pearson (1999).

[3] See, for example, Hull and White (1990a).

[4] See, for example, Kellerhals and Uhrig-Homburg (1998) for empirical results on the market of German government bonds.

The described difficulties with estimating term structure models have created a branch of econometric literature devoted to the analysis of panel data. This combination of time-series and cross-section data is considered very suitable for the interest rate market, since on every trading day a whole spectrum of interest rates, i.e. a term structure, is being provided from the markets of fixed-income instruments. An estimation based on interest rate panel data takes into account the whole information embedded in the term structure and should therefore result in more efficient parameter estimates. The approach we implement is based on a convenient state space representation of the term structure model and treats the underlying state variables correctly as unobservables. In conjunction with the customized state space model we then exploit the econometric methods of Kalman filtering and maximum likelihood estimation.

This estimation procedure is considered an appropriate method for both extracting time-series of the unobservable state variables and estimating the relevant model parameters.[5] The literature along this line can be traced back to the work by Pennacchi (1991) who estimates an equilibrium inflation term structure model with a number of four Treasury bills with different maturities. Ball and Torous (1996) and Duan and Simonato (1999) provide encouraging simulation results on Kalman filters working with single-factor models of the Vasicek (1977) and Cox, Ingersoll, and Ross (1985b) type. Especially, the mean-reversion parameters are estimated accurately when compared to other estimation procedures.[6] The work by Lund (1997) uses extended Kalman filters in order to use other marketable securities than pure yield curve data like, for example, coupon bond prices and swap rates. Further, Babbs and Nowman (1999) and De Jong (2000) demonstrate how to implement affine multi-factor term structure models on panel interest rate data. Nunes and Clewlow (1999) present estimation results on simulations and real market data for extended Kalman filter algorithms using interest rate caps and swaptions. Chen and Scott (2003) estimate one-, two-, and three-factor models of the Cox, Ingersoll, and Ross (1985b) type using a state space model in which estimates of the unobservable state variables are generated by a Kalman filter. Several tests are performed to examine whether these models can accurately capture the variability of the term structure over time. The paper by Cortazar,

[5] For further advantages see, for example, Geyer and Pichler (1999) and De Jong (2000).

[6] For an exploration of different estimation methods in the case of mean-reverting interest rate dynamics that are sufficiently close to a non-stationary process with a unit root see, for example, Ball and Torous (1996).

Naranjo, and Schwartz (2003) provides a modified standard Kalman filter approach to deal with missing observations in panel data which can be applied to a broad class of dynamic term structure models.

In the following sections we calibrate our term structure model to standard fixed-income instruments. For this purpose we customize the general state space model as presented in chapter 3 to reflect the time-series and cross-sectional properties of our theoretical term structure model at best and setup the Kalman filters along the lines of chapter 4. We implement two special types of Kalman filter algorithms, one for linear and one for non-linear functional relationships in the measurement equation, in order to calibrate our term structure model to the liquid US markets of Treasury Securities, LIBOR rates, and swaps.

12.2 Discrete Time Distribution of the State Variables

In setting up the specific state space model usable for Kalman filter algorithms, we begin with deriving the transition equation from the continuous time dynamics of the state variables. The aim for the transition equation is to represent the stochastic evolution of the state variables $\mathbf{x} = [r(t), \lambda(t)]'$ as good as possible. In section 9.3 we made the assumption of

$$d\mathbf{x} = \left[\boldsymbol{\mu}^{\mathbb{P}} + \mathbf{T}^{\mathbb{P}}\mathbf{x}\right] dt + \boldsymbol{\sigma} d\mathbf{W}_t^{\mathbb{P}} \tag{12.1}$$

for the underlying factors of our term structure model. This stochastic representation of the evolution of the state variables can be integrated to yield the solution[7]

$$\begin{aligned}\mathbf{x}_t &= e^{\mathbf{T}^{\mathbb{P}}(t-s)}\mathbf{x}_s + \int_s^t e^{\mathbf{T}^{\mathbb{P}}(t-\zeta)}\boldsymbol{\mu}^{\mathbb{P}} dt + \int_s^t e^{\mathbf{T}^{\mathbb{P}}(t-\zeta)}\boldsymbol{\sigma} d\mathbf{W}^{\mathbb{P}} \\ &= e^{\mathbf{T}^{\mathbb{P}}(t-s)}\mathbf{x}_s - \left(\mathbf{T}^{\mathbb{P}}\right)^{-1}\left(\mathbf{I} - e^{\mathbf{T}^{\mathbb{P}}(t-s)}\right)\boldsymbol{\mu}^{\mathbb{P}} + \int_s^t e^{\mathbf{T}^{\mathbb{P}}(t-\zeta)}\boldsymbol{\sigma} d\mathbf{W}^{\mathbb{P}}\end{aligned}$$

for $s < t$. From this almost explicit integral solution we obtain for the discrete time interval of $\Delta t = [s, t]$ the following *exact discrete time equivalent*:

[7] See, for example, Hirsch and Smale (1974, ch. 5).

$$\mathbf{x}_t = e^{\mathbf{T}^{\mathbb{P}}\Delta t}\mathbf{x}_{t-\Delta t} - \left(\mathbf{T}^{\mathbb{P}}\right)^{-1}\left(\mathbf{I} - e^{\mathbf{T}^{\mathbb{P}}\Delta t}\right)\boldsymbol{\mu}^{\mathbb{P}} + \boldsymbol{\eta}_t, \tag{12.2}$$

$$\text{with } \boldsymbol{\eta}_t = \int\limits_{t-\Delta t}^{t} e^{\mathbf{T}^{\mathbb{P}}(t-\zeta)}\boldsymbol{\sigma} d\mathbf{W}^{\mathbb{P}}.$$

Further, with the constant vector of the unconditional long-run means of the state variables $\mathbf{m}$ defined by[8]

$$\boldsymbol{\mu}^{\mathbb{P}} - \mathbf{T}^{\mathbb{P}}\mathbf{m} = 0,$$

i.e. the constant is given by $\mathbf{m} = -\left(\mathbf{T}^{\mathbb{P}}\right)^{-1}\boldsymbol{\mu}^{\mathbb{P}}$, we obtain for equation (12.2):

$$\begin{aligned}\mathbf{x}_t &= e^{\mathbf{T}^{\mathbb{P}}\Delta t}\mathbf{x}_{t-\Delta t} - \left(\left(\mathbf{T}^{\mathbb{P}}\right)^{-1}\mathbf{T}^{\mathbb{P}} - \left(\mathbf{T}^{\mathbb{P}}\right)^{-1}e^{\mathbf{T}^{\mathbb{P}}\Delta t}\mathbf{T}^{\mathbb{P}}\right)\mathbf{m} + \boldsymbol{\eta}_t \\ &= e^{\mathbf{T}^{\mathbb{P}}\Delta t}\mathbf{x}_{t-\Delta t} - \left(\mathbf{I} - e^{\mathbf{T}^{\mathbb{P}}\Delta t}\right)\mathbf{m} + \boldsymbol{\eta}_t\end{aligned}$$

for which we finally get

$$\mathbf{x}_t - \mathbf{m} = e^{\mathbf{T}^{\mathbb{P}}\Delta t}\left(\mathbf{x}_{t-\Delta t} - \mathbf{m}\right) + \boldsymbol{\eta}_t.$$

Thus, we derived our setting for the transition equation of the state space model as

$$\boldsymbol{\xi}_t = e^{\mathbf{T}^{\mathbb{P}}\Delta t}\boldsymbol{\xi}_{t-\Delta t} + \boldsymbol{\eta}_t. \tag{12.3}$$

using the substitution $\boldsymbol{\xi}_t = \mathbf{x}_t - \mathbf{m}$. For the remaining matrix exponential $e^{\mathbf{T}^{\mathbb{P}}\Delta t}$ of equation (12.3) we use diagonal Padé approximations with scaling and squaring defined as follows.[9]

Definition 12.1 (Padé Approximation) [10] *The (p, q) Padé approximation to $e^{\mathbf{A}}$, where $\mathbf{A}$ is a $n \times n$ matrix, is defined by*

$$\begin{aligned}R_{p,q}(\mathbf{A}) &= \left[D_{p,q}(\mathbf{A})\right]^{-1} N_{p,q}(\mathbf{A}), \\ \text{with } N_{p,q}(\mathbf{A}) &= \sum_{k=0}^{p} \frac{(p+q-k)!p!}{(p+q)!k!(p-k)!}\mathbf{A}^k, \text{ and} \\ D_{p,q}(\mathbf{A}) &= \sum_{k=0}^{q} \frac{(p+q-k)!q!}{(p+q)!k!(q-k)!}(-\mathbf{A})^k.\end{aligned}$$

---

[8] In section 12.3.3 we will see that m is the fix point of the deterministic part of the system described in equation (12.1).

[9] For a comparison in terms of computational stability of various algorithms to calculate matrix exponentials see Moler and van Loan (1978).

[10] See, for example, Golub and van Loan (1996).

Diagonal Padé Approximations are given if $p = q$. The method of scaling and squaring exploits the fundamental property

$$e^{\mathbf{A}} = \left(e^{\mathbf{A}/m}\right)^m$$

of the exponential function. With a j-th order scaling, i.e. choosing $m = 2^j$, we use

$$e^{\mathbf{A}} = \left[R_{q,q}\left(\frac{\mathbf{A}}{2^j}\right)\right]^{2^j}$$

as approximating algorithm to the matrix exponential.

Further, we need to examine the statistical properties of the noise term $\boldsymbol{\eta}_t$ in equation (12.3). For the conditional expectation of the error term we obtain $\mathbb{E}^{\mathbb{P}}\left[\boldsymbol{\eta}_t|\mathcal{F}_{t-\Delta t}\right] = \mathbf{0}$; its covariance matrix $\mathbf{Q}(\boldsymbol{\psi})$ will be approximated by

$$\begin{aligned}
\mathbb{E}^{\mathbb{P}}\left[\boldsymbol{\eta}_t\boldsymbol{\eta}_t'|\mathcal{F}_{t-\Delta t}\right] &= \mathbb{E}^{\mathbb{P}}\left[\left(\int\limits_{t-\Delta t}^{t} e^{\mathbf{T}^{\mathbb{P}}(t-\zeta)}\boldsymbol{\sigma}d\mathbf{W}\right)^2\middle|\mathcal{F}_{t-\Delta t}\right] \\
&= \mathbb{E}^{\mathbb{P}}\left[\int\limits_{t-\Delta t}^{t} e^{\mathbf{T}^{\mathbb{P}}(t-\zeta)}\begin{bmatrix}\sigma_r^2 & \rho\sigma_r\sigma_\lambda \\ \rho\sigma_r\sigma_\lambda & \sigma_\lambda^2\end{bmatrix}\right. \\
&\quad \left.\left(e^{\mathbf{T}^{\mathbb{P}}(t-\zeta)}\right)' dt\middle|\mathcal{F}_{t-\Delta t}\right] \approx \begin{bmatrix}\sigma_r^2 & \rho\sigma_r\sigma_\lambda \\ \rho\sigma_r\sigma_\lambda & \sigma_\lambda^2\end{bmatrix}\Delta t.
\end{aligned}$$

Bringing the results together, we can state the *transition equation* - as examined in general in chapter 2 - corresponding to the time-series implications of our term structure model by

$$\boldsymbol{\xi}_t = \mathbf{c}_t(\boldsymbol{\psi}) + \boldsymbol{\Phi}_t(\boldsymbol{\psi})\,\boldsymbol{\xi}_{t-1} + \boldsymbol{\eta}_t(\boldsymbol{\psi}), \tag{12.4}$$

$$\begin{aligned}
\text{with}\quad \mathbf{c}_t(\boldsymbol{\psi}) &= [0,0]', \\
\boldsymbol{\Phi}_t(\boldsymbol{\psi}) &= \left[R_{q,q}\left(\frac{e^{\mathbf{T}^{\mathbb{P}}\Delta t}}{2^k}\right)\right]^{2^k}, \\
\mathbb{E}\left[\boldsymbol{\eta}_t|\mathcal{F}_{t-\Delta t}\right] &= \mathbf{0}, \text{ and} \\
\mathbb{V}ar\left[\boldsymbol{\eta}_t|\mathcal{F}_{t-\Delta t}\right] &= \begin{bmatrix}\sigma_r^2 & \rho\sigma_r\sigma_\lambda \\ \rho\sigma_r\sigma_\lambda & \sigma_\lambda^2\end{bmatrix}\Delta t
\end{aligned}$$

for the matrix and error term specifications. The obtained transition equation can be used in our empirical models independently of the interest data we calibrate our term structure model to. However, the functional relationship of the measurement equation is conditional to the fixed-income instruments we use as sample data. Based on the sample data we use for calibration we need to differentiate between implementations of the linear and the extended Kalman filter algorithm as we show in the following two sections.

12.3 US Treasury Securities

12.3.1 Data Analysis

For a first calibration of our term structure model we employ the very liquid markets of US Treasury Securities. The database we use is set up from the Bliss (1999) data which is an updated version of the widely applied McCulloch-Kwon data.[11] The data set consists of month end price quotes for Treasury issues for the period January 1970 through December 1998 which yields 348 sample dates.

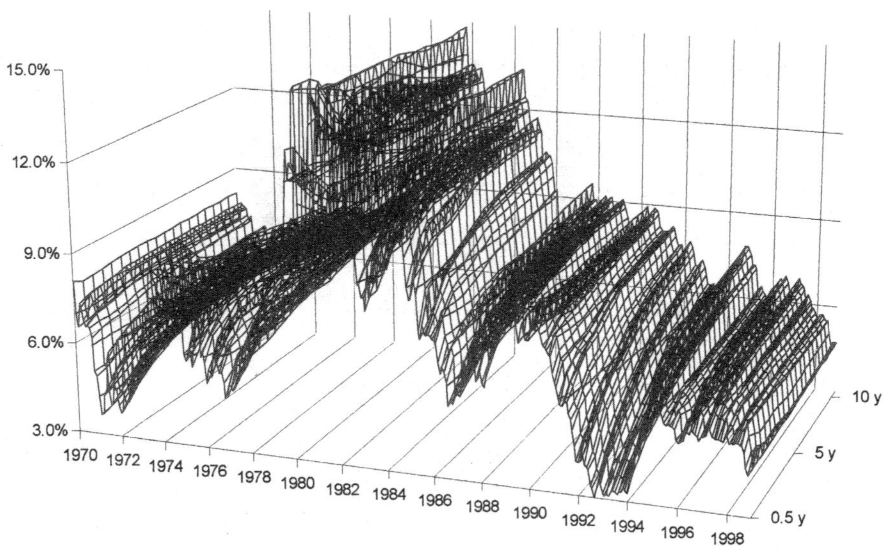

Fig. 12.1. Estimated Zero Bond Yield Curve from 1970 to 1998

The price quotes are taken from the *Center of Research in Security Prices* (CRSP) Government Bonds files. Included are all eligible issues with matu-

[11] For a detailed description of the data set see McCulloch and Kwon (1993).

rities running from 1 month to 30 years.[12] From these issues the continuous time yields are estimated by approximating the discount function by cubic splines.[13] This results in a full term structure of continuous time zero bond interest rates. In order to get an idea on how the term structure evolves over time we visualize the historical evolution of the interest rates in figure 12.1 and provide accompanying descriptive statistics in table 12.1. Looking at the shape of the term structure the data show an upward sloping yield curve on average. Moreover, we can identify the unstable period of the years 1979 to 1982 when the Federal Reserve monetary policy shifted to the 'New Operating Procedure'. The Federal Reserve Bank's anti inflation strategy from October 1979 to October 1982 was to deemphasize the federal funds rate as operating target and to turn to non-borrowed reserves targeting.[14] During this period high levels of interest rates coincide with large volatilities. The development of interest rates during this period is also the reason for the positive skewness of around one for the different maturities.

Table 12.1. Descriptive Statistics of the Term Structure Sample

Maturity	3 M	6 M	1 Y	2 Y	3 Y	5 Y	7 Y	10 Y
MEAN	0.0684	0.0709	0.0731	0.0761	0.0777	0.0802	0.0816	0.0828
STD	0.0271	0.0272	0.0263	0.0248	0.0237	0.0226	0.0220	0.0209
MAX	0.1619	0.1640	0.1621	0.1586	0.1573	0.1535	0.1507	0.1477
MED	0.0613	0.0644	0.0679	0.0702	0.0724	0.0749	0.0763	0.0776
MIN	0.0278	0.0288	0.0309	0.0380	0.0420	0.0435	0.0430	0.0451
SK	1.1801	1.1135	1.0447	1.0292	1.0448	1.0425	1.0247	1.0133
KU	4.3848	4.1363	3.9071	3.7443	3.6703	3.4985	3.3949	3.3956

Note:
For each maturity we report the sample mean (MEAN), standard deviation (STD), maximum values (MAX), median (MED), minimum values (MIN), skewness (SK), and kurtosis (KU).

In terms of maturities used for our state space model, i.e. for the observable number of yields $\mathbf{y}_t$ in the measurement equation (12.5), we need to choose a number of distinct maturities. For the choice of maturities we consider two relevant aspects: First, we need to constrain our measurement equation to a

[12] Bonds with option features and special liquidity problems were eliminated; see Bliss (1999).

[13] See McCulloch (1975) for the details of this method.

[14] For the Federal Reserve Bank's monetary operating procedures see, for example, Miller and VanHoose (1993, ch. 25).

fixed number of yields. Recent choices from the empirical literature on Kalman filtering cover the range from three yields as used in the simulation analysis of Ball and Torous (1996) to eight yields as with the empirical study by Babbs and Nowman (1999). In line with the works by Duan and Simonato (1999) and De Jong (2000) we consider a number of four yields as appropriate. Second, to cover the dynamics of the whole spectrum of yields we need to select the yields appropriately over the available maturities. At the long end of the term structure we only have a sparse number of available issues, especially in the 1970's. At the short end we only use Treasury bills that have maturities above one month to avoid including unexplainable price movements of issues close to their redemption time. For finally choosing the distance between the maturities we also take the volatility term structure into account.

From table 12.1 we infer that the longer interest rates are less volatile than the shorter rates. This calls for a closer representation of the shorter part of the term structure. The 10-years rate, for example, is only 77 percent as volatile as the 3-months rate. As we have seen in section 10.2.2 our interest rate model is capable to cover such volatility structures. Thus, we select the four maturities containing three months, one year, five years, and ten years to govern the measurement equation (12.5).

Table 12.2. Cross-Correlations among Interest Rates

Maturity	3 M	6 M	1 Y	2 Y	3 Y	5 Y	7 Y	10 Y
3 M	1							
6 M	0.9963	1						
1 Y	0.9849	0.9944	1					
2 Y	0.9584	0.9713	0.9890	1				
3 Y	0.9355	0.9500	0.9732	0.9959	1			
5 Y	0.8973	0.9128	0.9419	0.9789	0.9928	1		
7 Y	0.8713	0.8874	0.9196	0.9639	0.9826	0.9973	1	
10 Y	0.8484	0.8644	0.8985	0.9480	0.9702	0.9905	0.9972	1

Furthermore, we show the cross-correlation among the estimated yields in table 12.2 and figure 12.2. Surprising is the high degree of *comovements* among the different interest rates along the term structure. Even the short and the long end with the 3-months and 10-years rate exhibit a correlation of 0.8484. Especially the interest rates with adjacent time to maturities reveal an almost perfect correlation whereas the degree of correlation decreases with increasing difference in maturity. However, the amount of decorrelation inherent in the term structure intuitively displays the need to model the term structure not

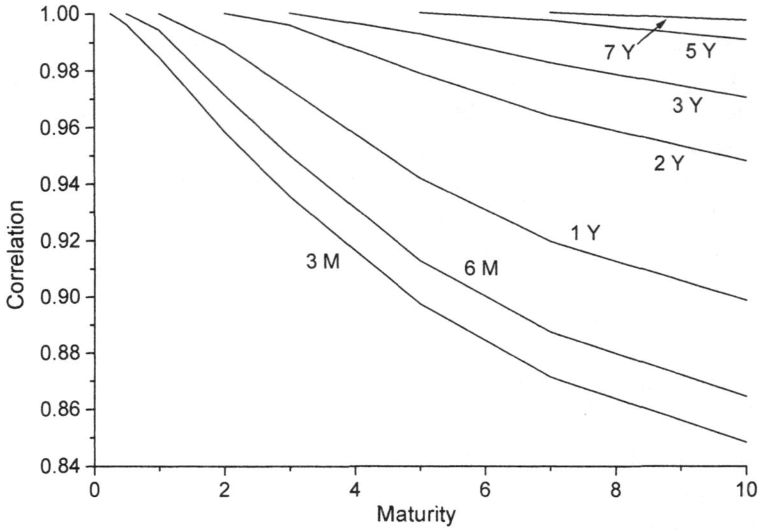

Fig. 12.2. Cross-Correlations among Interest Rates

only by a one but a multi-factor interest rate model of the yield curve. This stems from the fact that single-factor models imply perfect instantaneous correlation between the movements in yields of different maturities. However, this does not mean that with single-factor models the yield curve is forced to move in parallel shifts; the yields are affected by changes in the driving variable in as a complex way as the richness of the model can allow.[15]

The high correlations among the different interest rates give rise to the idea of analyzing the data set with the descriptive device of *principal component analysis*.[16] This analysis asks, whether we can describe each of a set of original variables by a linear function of a small number of so called common factors - the principal components - with a high degree of accuracy. This would be trivially true if all variables moved proportionally. Then a single variable would suffice to describe the behavior of all original variables. In general, the principal component technique transforms the original variables into orthogonal new variables called factors. By this transformation the total variance of the variables is preserved. Finally, the first principal component accounts for the maximum fraction of the total variance, the second factor explains the

[15] For the shortcomings of a single-factor model see, for example, Chen (1996b).

[16] This multivariate analysis technique is described in detail, for example, in Theil (1971, sec. 1.9) and Jolliffe (1986). For testing cointegration with principal component methods see, for example, Phillips and Ouliaris (1988).

largest part of the residual variance until the last factor accounts for all of the original variance.

In the case of a set of interest rates combined to build up a full spectrum of maturities, i.e. a yield curve, we face a high degree of comovement across the yields as described above.[17] The idea to determine common factors that affect interest rates of different maturities was first developed by Litterman and Scheinkman (1991). For US Treasury Securities their analysis suggests that most of the variation in returns on all fixed-income securities can be explained by three factors. A first factor is attributed to the level of interest rates in that it is basically constant across maturities. The second factor they call steepness factor even though it does not correspond exactly to any of the steepness measures commonly used. This factor lowers the yields with shorter maturities and raises the interest rates with higher dates of maturity, and vice versa. Finally, the last factor, which they call curvature, increases the curvature of the yield curve. The purpose of obtaining these independent factors is to be able to describe the dynamics of the entire yield curve without a large loss of information.

Based on our data set we further examine whether we are able to drastically reduce the number of independent factors with little loss of information. This will further support our proposed term structure model in that it only uses two factors, the instantaneous interest rate and the market price of risk. The results of the principal components analysis are presented in tables 12.3 and 12.4. The original variables are the returns of the different interest rates from the yield curve.[18]

Table 12.3. Principal Component Analysis

Principal	Component	First	Second	Third	Fourth	Fifth
Accounted	separate	0.871168	0.098036	0.017786	0.007466	0.002927
Variance	cumulated	0.871168	0.969204	0.986990	0.994456	0.997383

Table 12.3 shows the resulting values for the explained fractions of variance. The first line consists of the separate numbers of the accounted variance.

[17] For a discussion of cointegration in the case of the term structure of interest rates see, for example, Campbell and Shiller (1987). The formal development of the key concepts of cointegration is given in Engle and Granger (1987).

[18] For the principal component analysis we use all eight interest rates with maturities given in table 12.1.

Table 12.4. Factor Loadings of the Common Factors

Principal	Factor Loadings							
Component	3-M	6-M	1-Y	2-Y	3-Y	5-Y	7-Y	10-Y
First	0.1076	0.1126	0.1081	0.0973	0.0873	0.0731	0.0647	0.0570
Second	-0.0521	-0.0308	-0.0046	0.0144	0.0239	0.0302	0.0317	0.0320
Third	0.0159	-0.0015	-0.0145	-0.0139	-0.0093	0.0026	0.0123	0.0209
Fourth	0.0102	-0.0101	-0.0097	0.0039	0.0076	0.0083	0.0011	-0.0109
Fifth	-0.0038	0.0082	-0.0026	-0.0065	-0.0001	0.0067	0.0037	-0.0057

Therein, the first principal component already contains 87.1 percent of the total variance incorporated in the original variables. The second factor still has an explaining power of 9.8 percent. Thus, as can be seen from the second line, the first two factors add up to 96.9 percent of cumulated accounted variance. An additional third factor can raise the possible variance by another 1.8 percent, whereas the fourth and fifth factor only reveal another 0.7 and 0.3 percent, respectively. Thus, by implementing our two factor interest rate model we at best only loose 3.1 percent of the original variance inherent in the data sample.

Further, we present the factor loadings of each of the first five principal components in table 12.4. These weights represent the sensitivity of the original variables to the common factors. Thus, we can interpret the loadings as the impact of the factors on the different yields of the term structure. The results on the weights of the first three principal components are graphed in figure 12.3. For the first factor weights we obtain positive loadings in a fairly constant range of 0.1126 to 0.0570. This means that the yield changes caused by the first factor essentially represent a parallel change of the yield curve with the short maturities being somewhat more sensitive to the factor movement. This result gives evidence to the interpretation of the first principal component as being the average level of the yield curve. The second principal component is made up by continually increasing weights across the term structure. The factor loadings are of opposite signs at the short and long end of the maturity spectrum with a minimum at the 3-months yield of -0.0521 and a maximum of 0.0320 for the 10-years rate. Thereby, the second factor lowers the interest rates with shorter maturities and increases the longer maturity rates which leads to refer to this factor as steepness factor. Besides the first two principal components the remaining higher order factors exhibit significant lower weights. The highest weight of 0.0209 we obtain with the third factor. For this factor the loadings are positive at the very short end and for maturities above five years. In the intermediate maturity range we observe negative factor loadings. Thus, the third principal component makes the yield curve twist in the way that it raises it at the ends and at the same time lowers interest rates with intermediate maturities. The subsequent factors show even lower weights, at most half of the weights of the third factor, and do not exhibit distinct patterns in that we see theoretical interpretable facts. We conclude that for our data set we obtain generally comparable results to those presented in the study of Litterman and Scheinkman (1991).

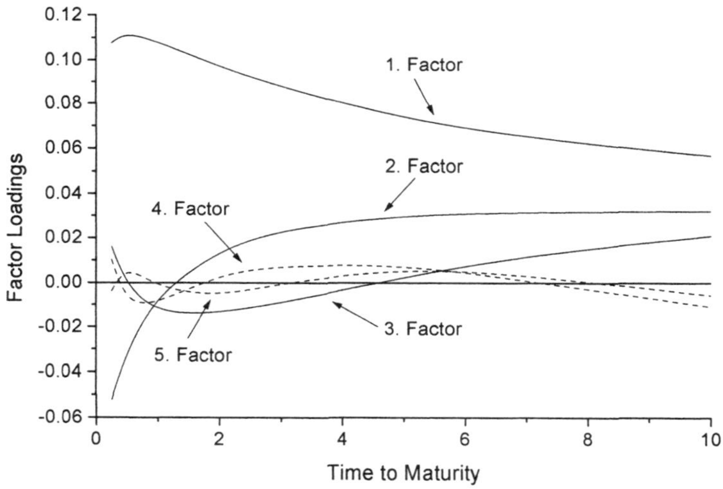

Fig. 12.3. Factor Loadings of Principal Components

In the light of these findings, the reduction of modeling the yield curve with only specifying the dynamics of two factors, as with our theoretical term structure model, is further given ground from a statistical analysis of the historical interest rate data. Next, we present the estimation results on the parameter values of our term structure model before we turn to analyze the behavior of the underlying state variables.

12.3.2 Parameter Estimation

In the case of extracting the information on interest rates from zero coupon spot rates, we are able to set up a linear Kalman filter algorithm as laid out in chapter 4. In order to obtain a measurement equation we exploit the affine functional relationship

$$y(t,T) = \frac{1}{T-t}\left[\left[A(t,T), B(t,T)\right]\mathbf{x} + C(t,T)\right]$$

from section 10.2.1. Restating this affine structure for a vector of g zero bond yields $\mathbf{y}(t,T)$ and the driving factors, the short rate $r(t)$ and the market price of risk $\lambda(t)$, yields the *measurement equation*

$$\mathbf{y}_t = \mathbf{a}_t(\psi) + \mathbf{B}_t(\psi)\boldsymbol{\xi}_t + \varepsilon_t(\psi), \tag{12.5}$$

$$\begin{aligned} \text{with } \mathbf{a}_t(\psi) &= [\mathbf{A}^*, \mathbf{B}^*]\,\mathbf{m} + \mathbf{C}^*, \text{ and} \\ \mathbf{B}_t(\psi) &= [\mathbf{A}^*, \mathbf{B}^*], \end{aligned}$$

where the modified factor loadings are given by $\mathbf{O}^* = [O(t,T_1)/(T_1-t), O(t,T_2)/(T_2-t), \ldots, O(t,T_g)/(T_g-t)]'$ with $\mathbf{O}^* \in \{\mathbf{A}^*, \mathbf{B}^*, \mathbf{C}^*\}$. In the measurement equation we further add a normally distributed measurement error term $\varepsilon_t(\psi)$ for that the observable yields $\mathbf{y}_t$ are only measurable with noise. For the moment specification of the error term $\varepsilon_t(\psi)$ we assume that the errors have zero mean, are serially uncorrelated, and have a time invariant covariance matrix. Notationally, we have the assumptions

$$\begin{aligned} \mathbb{E}\left[\varepsilon_t|\mathcal{F}_{t-\Delta t}\right] &= \mathbf{0}, \text{ and} \\ \mathbb{V}ar\left[\varepsilon_t|\mathcal{F}_{t-\Delta t}\right] &= \sigma_\varepsilon^2 \cdot \mathbf{I}_{g\times g} \end{aligned}$$

which adds with σ_ε one more parameter to our parameter space ψ.[19]

Thus, with the functional mapping of the state variables $\boldsymbol{\xi}_t$ on the observable yield vector $\mathbf{y}_t$ from the bond market, we establish our specific measurement equation. Based on the *state space model* defined by the transition equation (12.4) and the measurement equation (12.5) we are able to run a linear Kalman filter algorithm. Applying the filter algorithm we estimate the parameter values ψ as laid out in chapter 5 by means of maximum likelihood. Based on the term structure data set of US Treasury Security issues we estimate the parameter values for the full available period ranging from 1970 to 1998.

In table 12.5 we present the results for the entire sample period of our two-factor model as well as of the reduced model which we derived in section

[19] Note that this homoskedastic error term specification leaves space to incorporate the full cross-correlation structure among interest rates of different maturities as given in equation (10.22) in order to obtain a close mapping with the empirical values reported in table 12.2.

Table 12.5. Estimated Parameter Values

Parameters	Our Model		Reduced Model	
ψ	EST	STD	EST	STD
μ_r	0.007068**	0.003397	0.004255*	0.002253
κ_r	-0.074414***	0.008277	-0.067469***	0.004440
σ_r	0.022532***	0.000926	0.017075***	0.000992
μ_λ	0.000768	0.122034	-0.255242**	0.131353
κ_λ	-0.619839***	0.120246		
σ_λ	0.386329***	0.022684		
θ_λ	-0.443361***	0.019980		
λ_2	-0.116206	0.172479		
ρ	0.710261***	0.033481		
σ_ε	0.001400***	0.000036	0.006011***	0.000125
$L(\mathbf{y};\psi)$	-6844.58		-4916.87	

Notes:
Statistically significant parameter estimates at the 1%-, 5%- and 10%-levels are denoted by ***, **, and *, respectively. The parameter μ_λ corresponds to the constant λ in the reduced model.

10.3.1. For the parameters common to both models we come to comparable estimates besides for parameters μ_λ and σ_ε. The parameter μ_λ actually corresponds to the constant market price of risk λ in the reduced model. With a value of -0.2552 the constant value actually lies close to the average value of the stochastic market price of risk in our model which is given in table 12.7 on page 163 with a mean value of -0.2624. Looking at the values of the measurement error term σ_ε we see a significant improvement by using the two-factor model. Adding one more factor lowers the error term from 60 b.p. to 14 b.p. Overall, the parameter estimates are statistically significant besides the two level parameters of the market price of risk process, μ_λ and λ_2. This, however, is not surprising since drift parameters generally show wide confidence intervals. The mean-reversion parameters κ_r and κ_λ imply mean half-lives, i.e. the expected time for the state variables to return halfway to their long-run means, of 9.3 years and 1.1 years, respectively. These results correspond to the estimates on the volatility parameters in that $\sigma_r = 2.25\%$ is much lower than the volatility of the market price of risk with $\sigma_\lambda = 38.6\%$. The point estimates on the parameters μ_r, κ_r, σ_r, and σ_ε, which are comparable to the one and two-factor specifications of term structure models tested by Duan and Simonato (1999) and Babbs and Nowman (1999), are found to be in line with their reported findings. For the correlation factor ρ in our model we obtain a point estimate of 0.71. This demonstrates that in times of high short rates the market price of short rate risk λ is also high which means that market participants then parallel become more risk taking.[20]

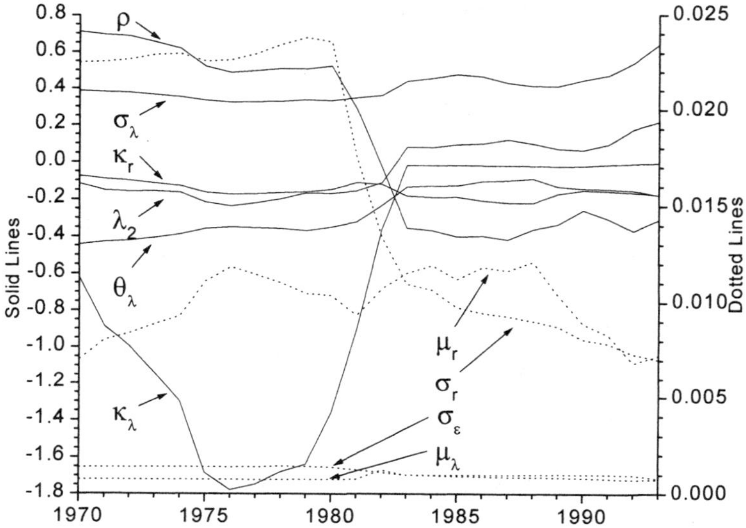

Fig. 12.4. Estimated Values for the Parameters ψ over Time

Further, we examine the stability of the parameter estimates for different choices of the sample period. In detail we run inferences for the whole sample period from 1970 to 1998 and continuously decrease the sample length by one year each time ending in a minimum sample of five years from 1993 to 1998. The results on these fourteen inferences are shown graphically in figure 12.4 and the accompanying descriptive statistics are reported in table 12.6. At a first glance of the graph we see that the point estimates of the parameters are fairly stable until 1979 and also for the period from 1982 onwards. In the previous section we saw that interest rates follow a different pattern during the late seventies and early eighties due to the Federal Reserve Bank's changing policy. Our results support the finding in the literature that the shift in the monetary policy during the years of 1979 to 1982 caused a structural break in

[20] For the impact of the state variables on the term structure compare section 10.4.1.

Table 12.6. Statistics on Historical Parameter Estimates

Parameter	MEAN	STD	MAX	UQ[a)]	MED[a)]	LQ[a)]	MIN	SK[b)]	KU[b)]
μ_r	0.0100	0.0017	0.0121	0.0115	0.0104	0.0088	0.0069	-0.48	1.90
κ_r	-0.1577	0.0396	-0.0744	0.0097	0.0090	0.0083	-0.2222	0.37	2.45
σ_r	0.0159	0.0070	0.0238	0.0085	0.0100	0.0118	0.0070	-0.02	1.14
μ_λ	0.0008	0.0001	0.0013	0.0120	0.0087	0.0117	0.0007	2.10	7.67
κ_λ	-0.6801	0.7048	-0.0109	0.0119	0.0089	0.0096	-1.7787	-0.37	1.50
σ_λ	0.4049	0.0766	0.6373	0.0097	0.0076	0.0096	0.3238	1.27	4.63
θ_λ	-0.2608	0.1275	-0.0918	0.0113	0.0100	0.0109	-0.4434	-0.01	1.29
λ_2	-0.0442	0.1473	0.2139	0.0105	0.0089	0.0096	-0.2379	0.21	1.43
ρ	0.1145	0.4652	0.7103	0.0109	0.0100	0.0117	-0.4187	0.03	1.16
σ_ε	0.0012	0.0002	0.0015	0.0117	0.0100	0.0117	0.0008	-0.14	1.41

Notes:
a) In the statistics for the parameter estimates, UQ and LQ denote the upper and lower quartile respectively; the median is abbreviated by MED.
b) For each parameter we also report the skewness (SK) and kurtosis (KU).

the interest rate process.[21] This can especially be seen from the volatility of the short rate σ_r which starts at 225 b.p. fairly constant until 1980 where it falls monotonically to a value of 7 b.p. for the minimum sample length. Therein, the impact of the high volatilities during these years becomes evident. Also, the correlation factor between the two state variables changes from positive values of 0.7 to -0.3 which marks a shift in the market participants changes of risk aversion attitudes. Further, during time the mean-reversion tendency of the market price of risk with a maximum of -1.78 does not become significant any more after 1982 whereas the central tendency parameter of the short rate κ_r rather stays constant with maximum and minimum values of -0.07 and -0.22, respectively. Finally, we see the measurement error σ_ε to become smaller with a decreasing sample size starting from a value of 14 b.p. to end with values 8 b.p. This is a reasonable fact for the parameter σ_ε since it is a parameter to measure the goodness of fit which needs to get better for a smaller sample size.

[21] See, for example, Duan and Simonato (1999) and Hansen (1998).

12.3.3 Analysis of the State Variables

In the analysis of the state variables we first examine the dynamic behavior of the two-factor term structure model in the absence of randomness before we examine the obtained time-series of the state variables from the sample data.

The description of the dynamic behavior of the state variables can either be based on equation (9.1) or equation (9.2) dependent on whether we look at the dynamics under the $\mathbb{P}$-measure or the $\mathbb{Q}$-measure. Without stochastic randomness we obtain the system

$$\frac{d\mathbf{x}^{\mathbb{M}}}{dt} = \boldsymbol{\mu}^{\mathbb{M}} + \mathbf{T}^{\mathbb{M}}\mathbf{x}^{\mathbb{M}} \tag{12.6}$$

for the state variables $\mathbf{x}$ under measure $\mathbb{M} \in \{\mathbb{P}, \mathbb{Q}\}$. Choosing the market price of risk as the dependent state variable we divide the given equations by each other which results in the differential equations

$$\frac{d\lambda^{\mathbb{P}}}{dr^{\mathbb{P}}} = \frac{\mu_\lambda + \kappa_\lambda r^{\mathbb{P}} + \theta_\lambda \lambda^{\mathbb{P}}}{\mu_r + \kappa_r r^{\mathbb{P}}}, \text{ and } \frac{d\lambda^{\mathbb{Q}}}{dr^{\mathbb{Q}}} = \frac{\mu_\lambda - \sigma_\lambda \lambda_2 + \kappa_\lambda r^{\mathbb{Q}} + \theta_\lambda \lambda^{\mathbb{Q}}}{\mu_r + \kappa_r r^{\mathbb{Q}} - \sigma_r \lambda^{\mathbb{Q}}}.$$

These equations define the *phase space* of the dependent state variable λ. The solutions to these first order differential equations can be plotted in a *vector field* as graphed in figure 12.5.[22] The direction, the angle, and the length of the vector line segments represent, respectively, the movement, the way, and the speed of the state variables from the starting values at that point on an orbital trajectory.

In order to algebraically analyze the *stability*[23] of our state space system we first calculate the eigenvalues of the system as

$$ew_{\mathbb{P}} = \{-0.4434; -0.0744\}, \text{ and } ew_{\mathbb{Q}} = \{-0.4797; -0.0398\}$$

with corresponding eigenvectors of

$$ev_{\mathbb{P}} = \left\{ \begin{bmatrix} 0 \\ 1 \end{bmatrix}; \begin{bmatrix} -0.5115 \\ 0.8593 \end{bmatrix} \right\}, \text{ and } ev_{\mathbb{Q}} = \left\{ \begin{bmatrix} 0.0557 \\ 0.9984 \end{bmatrix}; \begin{bmatrix} -0.5456 \\ 0.8380 \end{bmatrix} \right\}.$$

Under both measures we obtain eigenvalues with the properties[24]

[22] The vector field is based on the estimated parameter values reported in table 12.5.

[23] For a systematic classification of the possible equilibria of a two dimensional linear system see, for example, Brock and Malliaris (1989, ch. 3.6).

[24] See, for example, Brock and Malliaris (1989, p. 80).

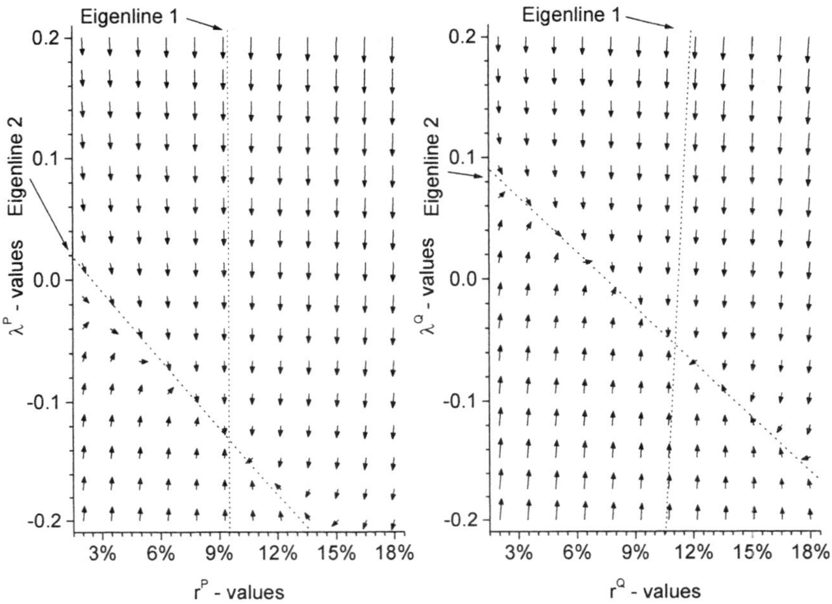

Fig. 12.5. Phase Plane of State Variables **x** as Vector Field

$$ew_1 < ew_2 < 0,$$

i.e. they are distinct and lie both below zero. In this case we obtain a *stable tangent node* at $\mathbf{x}^{\mathbb{P}} = [0.0950, -0.1311]'$ and $\mathbf{x}^{\mathbb{Q}} = [0.1106, -0.0517]'$, respectively, as indicated by the point of intersection of the two dotted lines.[25] The dotted lines represent the *eigenlines* of the system which are lines through the asymptotic point of the system. They can be derived from the general solutions to the partial differential equation of (12.6)

$$\begin{bmatrix} r^{\mathbb{P}} \\ \lambda^{\mathbb{P}} \end{bmatrix} = \begin{bmatrix} 0.0950 \\ -0.1311 \end{bmatrix} + c_1 \begin{bmatrix} 0 \\ 1 \end{bmatrix} e^{-0.4434} + c_2 \begin{bmatrix} -0.5115 \\ 0.8593 \end{bmatrix} e^{-0.0744}, \text{ and}$$

$$\begin{bmatrix} r^{\mathbb{Q}} \\ \lambda^{\mathbb{Q}} \end{bmatrix} = \begin{bmatrix} 0.1106 \\ -0.0517 \end{bmatrix} + c_3 \begin{bmatrix} 0.0557 \\ 0.9984 \end{bmatrix} e^{-0.4797} + c_4 \begin{bmatrix} -0.5456 \\ 0.8380 \end{bmatrix} e^{-0.0398},$$

with arbitrary constants c_i, with $i \in \{1, 2, \ldots, 4\}$. The eigenlines are found by consecutively setting the constants c_i equal to zero and solving the parametric representation of the functional relationships between r and λ for the state

[25] The coordinates of the nodes are found by setting the first order derivatives of equation (12.6) equal to zero and then solve for the state variables.

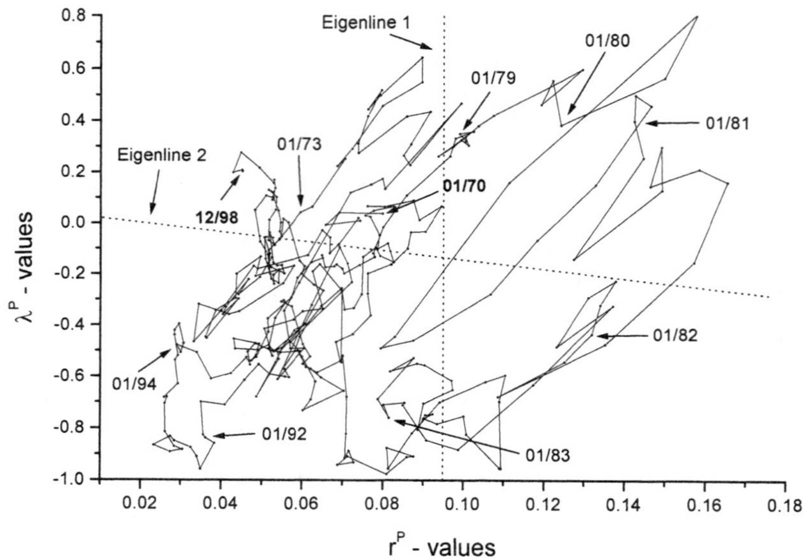

Fig. 12.6. Historical Evolution of State Variables from 1970 to 1998

variable λ. Generated by the eigenvectors $[0,1]'$ and $[-0.5115, 0.8593]'$ for the dynamic system under the measure $\mathbb{P}$ we obtain

$$r_1^{\mathbb{P}} = 0.0950, \text{ and } \lambda_2^{\mathbb{P}} = -1.6800 r_2^{\mathbb{P}} + 0.0285$$

for the eigenlines. For the dynamic system under $\mathbb{Q}$ the eigenlines become

$$\lambda_1^{\mathbb{Q}} = 17.9100 r_1^{\mathbb{Q}} - 2.0332, \text{ and } \lambda_2^{\mathbb{Q}} = -1.5359 r_2^{\mathbb{Q}} + 0.1182$$

for the eigenvectors $[0.0557, 0.9984]'$ and $[-0.5456, 0.8380]'$, respectively. Both eigenlines are plotted in dashed lines for the two probability measures in figure 12.5. From the vector line segments we see that the dynamic system is stable. Starting from any point in the phase space the path is forced on different ways at a decreasing speed towards the asymptotic eigenlines and finally towards the long-term equilibrium state of the system.

In the analysis up to now we have eliminated the randomness of the dynamic system of the two modeled state variables. Now, we consider the dynamic behavior of the term structure model including random shocks. We especially examine the time-series for the unobservable state variables $\mathbf{x}_t$ retrieved from the data sample using the Kalman filter algorithm. In figure 12.6 we plot the historical evolution of the state variables $\mathbf{x}_t$ obtained from the filter algorithm for the whole sample period; included are the asymptotic lines from the deterministic analysis. The accompanying descriptive statistics are reported in table 12.7.

Table 12.7. Statistics on Realized State Variable Values

$\mathbf{x}$	MEAN	STD	MAX	UQ$^{a)}$	MED$^{a)}$	LQ$^{a)}$	MIN	SK$^{b)}$	KU$^{b)}$
r	0.0699	0.0280	0.1654	0.0826	0.0631	0.0519	0.0233	1.06	4.10
λ	-0.2624	0.3777	0.8084	0.0099	-0.2782	-0.5301	-0.9772	0.23	2.41

Notes:
a) In the statistics for the state variables $\mathbf{x}$, UQ and LQ denote the upper and lower quartile respectively; the median is abbreviated by MED.
b) For each state variable we also report the skewness (SK) and kurtosis (KU).

From the numerical results we first see the short rate in a range of realized values from a minimum of 2.3% to a maximum of 16.5% with a mean value of 7.0%. These values closely resemble the statistics of the short rate (3-months rate) included in the data sample as can be seen from the reported values in table 12.1. Also, the positive skewness stemming from the 1979-1982 period (see the short rate values in graph 12.6) and the excess kurtosis are retained. Second, with our filtering technique we can extract the market prices of risk over time. The values we obtain for λ_t take a minimum of -0.98, a mean value of -0.26, and a maximum of 0.81. With an upper quartile value of close to zero we can infer that the market participants were risk averse over 75 percent of the sample period from 01/1970 - 12/1998. Finally, we see from graph 12.6 how the two state variables evolve over time. Therein, the years of 1979 to 1982 mark a period of high short rates along with risk taking values for the market price of risk. We further note that the path of the state variables actually crosses the asymptotic lines when we include the historical randomness. The central tendency is especially effective when the realized values are far away from their long-run means.

12.4 Other Liquid Markets

The analysis of the previous section is based on interest rates obtained from the US Treasury Security market. Other possible observables from the fixed-income markets include, for example, LIBOR or swap rates and prices for bond options, cap and floor agreements. Using such fixed-income instruments give rise to non-linear functional relationships between the state variables of our term structure model and the observable data. In thus, we further implement extended Kalman filter algorithms instead of the linear algorithms we could use for the US Treasury Security market. From the possible market prices of fixed-income instruments we choose to work with data from the LIBOR and swap markets because of their availability and high liquidity.[26]

12.4.1 Appropriate Filtering Algorithm

In order to calibrate our term structure model to the LIBOR and swap markets we implement an extended Kalman filter algorithm. The evolution of the state variables over time stays unchanged in that we can keep the specification of the transition equation (12.4) in the state space setting. However, the measurement equation needs to be linearized when we use LIBOR and swap rates for the observable vector $\mathbf{y}_t$ as we show in the following. Using our closed-form solution to the discount bond price from equation (10.16) the LIBOR and swap rates are given by the formulae

$$L(t,T) = \frac{1 - P(t,T)}{(T-t)\,P(t,T)}, \quad \text{and} \tag{12.7}$$

$$k(t,T) = \frac{1 - P(t,T)}{\delta \sum_{j=1}^{n} P(t,T_j)} \tag{12.8}$$

as defined in equations (8.3) and (11.16). From these formulae we see that the LIBOR curve represents the natural extension of the swap rate curve for shorter maturities for which no swap contracts are traded.[27] Thus, in order to cover the whole spectrum of the term structure we choose LIBOR and swap rates as observables $O(t,T_i) \in \{L(t,T_i), k(t,T_i)\}$ with which we specify the non-linear measurement equation as

$$\begin{aligned} \mathbf{g}_t\left(\boldsymbol{\xi}_t, \boldsymbol{\varepsilon}_t(\boldsymbol{\psi}), \boldsymbol{\psi}\right) &= \mathbf{s}\left(\boldsymbol{\xi}_t, \boldsymbol{\psi}\right) + \boldsymbol{\varepsilon}_t(\boldsymbol{\psi}), \\ \text{with } \mathbf{s}\left(\boldsymbol{\xi}_t, \boldsymbol{\psi}\right) &= \left[O(t,T_1), O(t,T_2), \ldots, O(t,T_g)\right]'. \end{aligned}$$

This non-linear equation can be linearized with a Taylor series extension as described in section 2.2.4. The first order approximations lead to the linearized

[26] Liquidity in the US interest rate swap market is generally agreed to be due to the high demand on swaps and the ease of hedging swaps with the US Treasury and futures markets. Also standard terms introduced by the ISDA and BBA in 1985 have assisted the growth in the swap markets. See, for example, Das (2004).

[27] The similarity of the two rates can be traced back to definition 11.6 where the floating part of the interest rate swap is defined on LIBOR.

form of the *measurement equation*

$$\mathbf{y}_t \approx \mathbf{a}_t + \mathbf{B}_{t|t-1}\boldsymbol{\xi}_t + \boldsymbol{\varepsilon}_t, \tag{12.9}$$

$$\text{with } \mathbf{a}_t = \mathbf{s}_t\left(\boldsymbol{\xi}_{t|t-1} + \mathbf{m}, \boldsymbol{\psi}\right) - \mathbf{B}_{t|t-1}\boldsymbol{\xi}_{t|t-1}, \text{ and}$$

$$\mathbf{B}_{t|t-1} = \left.\frac{\partial \mathbf{s}_t(\boldsymbol{\xi}_t, \boldsymbol{\psi})}{\partial \boldsymbol{\xi}_t'}\right|_{(\boldsymbol{\xi}_t, \boldsymbol{\psi}) = (\boldsymbol{\xi}_{t|t-1} + \mathbf{m}, \boldsymbol{\psi})}$$

in the case of the extended Kalman filter. With the discount bonds $P(t,T)$ and $P(t,T_j)$ being a function of the state variables the derivatives for the LIBOR and the swap rate can be calculated from equations (12.7) and (12.8) in closed-form. The moment assumptions regarding the error term $\boldsymbol{\varepsilon}_t(\boldsymbol{\psi})$ are identical with those in the linear case. With equation (12.9) we obtain a linearized version of the non-linear functional relationship between the unobservable state variables $\boldsymbol{\xi}_t$ and the LIBOR and swap rates $\mathbf{y}_t$ which are observable in the financial markets.

In analogy to the linear case, we are now able to run the extended Kalman filter algorithm of section 2.2.4 based on the *state space model* defined by the transition equation (12.4) and the measurement equation (12.9). The empirical implementation and the estimation results are presented in the next section.

12.4.2 Sample Data and Estimation Results

In this section we examine other liquid US fixed-income markets besides the US Treasury Security market underlying the analysis of section 12.3. With the specific form of the extended Kalman filter algorithm described in the previous section we are able to implement our term structure model on LIBOR and swap rates. The calibration procedure is based on interest rate data sampled monthly over the longest available period from 04/1987 to 01/2000. We obtain 154 observation dates and at each date we use four distinct maturities as in the case of the Treasury Security sample. In both, the LIBOR and the swap market, we select the maturities with the aim of capturing a maximum spectrum of the available term to maturities. For the LIBOR rates we choose to include the 1-, 3-, 6-, and 12-months rates and for the swap rates the maturities of 2, 3, 5, and 10 years. The closer maturity choices at the short end are intended to better capture the information of this more volatile part of the term structure.

In table 12.8 we report the summary statistics separately for each time-series. For the short-term market of LIBOR rates the values of the statistics for the different maturities show very similar results. The term structure rises on

Table 12.8. Descriptive Statistics of LIBOR and Swap Data

	LIBOR Data				Swap Data			
Maturity	1-M	3-M	6-M	12-M	2-Y	3-Y	5-Y	10-Y
MEAN	0.0598	0.0607	0.0617	0.0639	0.0672	0.0698	0.0731	0.0771
STD	0.0175	0.0174	0.0172	0.0171	0.0163	0.0156	0.0147	0.0141
MAX	0.1006	0.1025	0.1050	0.1094	0.1064	0.1042	0.1022	0.1070
MED	0.0569	0.0576	0.0586	0.0602	0.0632	0.0653	0.0697	0.0733
MIN	0.0313	0.0319	0.0325	0.0338	0.0391	0.0434	0.0493	0.0518
SK	0.3406	0.3224	0.3212	0.3288	0.3148	0.3240	0.3147	0.2733
KU	2.5095	2.4863	2.4952	2.4996	2.1952	2.0009	1.8238	1.7921

Note:
For each data series we report the sample mean (MEAN), standard deviation (STD), maximum values (MAX), median (MED), minimum values (MIN), skewness (SK), and kurtosis (KU).

Table 12.9. Parameter Estimates for LIBOR and Swap Data

Parameter	LIBOR Data		Swap Data		Combined Data	
ψ	EST	STD	EST	STD	EST	STD
μ_r	0.0174***	0.0045	0.0029***	0.0001	0.0041***	0.0009
κ_r	-0.2772***	0.0751	-0.0453***	0.0004	-0.0424***	0.0117
σ_r	0.0100***	0.0006	0.0156***	0.0003	0.0100***	0.0008
μ_λ	0.0113	0.0199	0.0037***	0.0003	0.0060***	0.0010
κ_λ	-0.5956	0.6618	-0.6168***	0.0044	-0.6925***	0.2204
σ_λ	3.2225***	0.3644	0.3811***	0.0132	1.3362***	0.2231
θ_λ	-1.8756***	0.2707	-0.4581***	0.0137	-0.8193***	0.1507
λ_2	0.1847	0.1532	0.0224***	0.0006	0.0196*	0.0109
ρ	0.1365	0.1255	0.6363***	0.0218	0.5566***	0.0745
σ_ε	0.0007***	0.0000	0.0003***	0.0000	0.0020***	0.0001
$L(\mathbf{y};\psi)$	-3393.86		-3674.51		-2969.80	

Note:
Statistically significant parameter estimates at the 1%-, 5%- and 10%-levels are denoted by ***, **, and *, respectively.

average with mean sample values of 5.98% for the one-month rate up to 6.39% for the 12-months rate. The volatility of the rates slightly decreases with higher maturities. For the criteria of skewness and kurtosis we see modest positively skewed rates that do not show excess kurtosis. These sample facts mostly resemble those for the swap rates. Due to the broader maturity spectrum we only see the term structure of swap rates increasing up to a 10-years rate of 7.71%.

The estimation results on the parameter values ψ of our term structure model are given in table 12.9. We calibrate the term structure model to three different panel data: The four LIBOR rates, the four swap rates, and a combination of the 1- and 12-months LIBOR rates along with the 5- and 10-years swap rates. The combined data set is motivated by the close relation of the two types of interest rate markets as we notice from equations (12.7) and (12.8) with the aim of obtaining a smooth term structure pattern over all available maturities. First, we compare the model fit to the three different samples in terms of the measurement error. The LIBOR rates exhibit a standard deviation of 7 b.p. which falls to 3 b.p. for the swap sample but increases to 20 b.p. for the combined data. This indicates that the swap market is the most homogeneous sample despite its high spanning of maturities from 2 to 10 years. Second, the results on the point estimates of μ_r, κ_r, and σ_r are comparable among the three samples in use. Third, the parameters corresponding to the stochastic specification of the market price of short rate risk are estimated with plausible values for the swap and the combined data. However, based on the LIBOR data we obtain diverging estimates for the parameters μ_λ, κ_λ, σ_λ, θ_λ and λ_2. This provides evidence that a sample including only a small one year window of the term structure is not a rich enough sample in order to get reliable estimates of the market price of risk parameters. Fourth, the mean-reversion parameters κ_r and κ_λ estimated for the swap and combined data at around -0.045 and -0.65 imply mean half-lives of 15.4 years and 1.1 years, respectively. Finally, the correlation coefficient is estimated at significantly positive values of around $+0.6$. Overall, all parameters are estimated at statistically significant levels besides the estimates for μ_λ, κ_λ, λ_2, and ρ based on the LIBOR data and λ_2 for the combined data.

13

Summary and Conclusions

In part III of this book we present a new continuous time term structure model and provide its implications on risk management and the pricing of different fixed-income derivatives based on a model calibration to standard fixed-income instruments. The theoretical and empirical analysis contributes to the term structure literature in that it shifts the theoretical attention of modeling static investor tastes towards capturing beliefs of investor risk attitudes towards interest rate risk using an affine model structure.

In the introductory chapter we consider the specific financial instruments most relevant in modeling fixed-income markets and describe the peculiarities of an incomplete market such as the market for fixed-income instruments. In the following chapter 9 we motivate our desire to build a new term structure model that takes into account time varying market prices of interest rate risk. The model we suggest chooses the short interest rate and its market price of risk as the two explaining state variables whereupon we establish an affine term structure model. In chapter 10 we present initial characteristic results and implications of the proposed term structure model. First, we derive a closed-form solution for arbitrage-free prices of discount bonds. Thereupon, we are able to state and analyze the implied term structures of interest rates and volatilities. In a comparative analysis we examine limiting cases of the term structures and the influence of the state variables and the model parameters on the shape of the term structures. We identify the first model factor as a level factor and the market price of risk as steepness factor that captures at the same time the investors' risk attitudes. Chapter 11 shows how to implement risk management techniques based on the proposed model. With the duration concept we illustrate the implementation of a simple model for interest rate risk management. More complex risk management strategies can be followed by using the derived closed-form expressions for the arbitrage-free prices of bond options, swap contracts, and cap and floor agreements.

In chapter 12 we show how to calibrate our term structure model to the standard fixed-income instruments of US Treasury Securities, LIBOR rates, and swap rates. The maximum likelihood estimation technique we implement is based on a linear as well as an extended Kalman filter algorithm. The parameter estimates and the data fit we obtain are similar to the reported values of comparable models in the literature. With the filtering technique we are able to extract the dynamic behavior of the risk attitudes of fixed-income investors. We infer that the market participants were risk averse over 75 percent of the sample period from 01/1970 - 12/1998 with a market price of risk ranging from a minimum of -0.98 to a maximum of 0.81 with a mean value of -0.26. From the historical evolution of the state variables over time we see that the years of 1979 to 1982 mark a period of high short rates along with risk taking values for the market price of risk. Finally, we report the estimation results on the liquid markets of US LIBOR and swap rates which are comparable to those obtained on the Treasury sample.

Part IV

Pricing Electricity Forwards

14

Introduction and Survey

14.1 Overview

In part IV we develop a continuous-time pricing model for one of the latest cash markets to be transformed by derivative securities - the electricity market. Based on available data on electricity derivatives we focus our attention on the valuation of short-term electricity forwards. First, we investigate the more mature futures markets contracted on established commodity underlyings and describe proposed pricing models for traditional commodity futures contracts from the literature. Thereby, we gather relevant information about the commodity electricity in order to build an appropriate valuation model to price electricity forwards in chapter 15. From given model assumptions we derive a closed-form valuation model for electricity forwards using the risk-neutral pricing technique of the EMM-approach. The suggested model especially captures the high and time varying volatility seen in electricity prices. In chapter 16 we present an empirical adaptation of the theoretical pricing model in state space form. Using maximum likelihood estimation based on extended Kalman filtering we report empirical results on electricity data from the largely deregulated Californian electricity market. In the last chapter we conclude.

14.2 Commodity Futures Markets

Basically, a futures contract is an agreement between two parties to make an exchange of financial products or commodities at a particular future date. Futures contracts are traded on futures exchanges which evolved from entrepreneurs pursuing an economic advantage by organizing the forward exchange in a standardized and regulated manner. With the evident risk of fluctuating commodity prices in the grain market of Chicago during the 18th century a group of businessmen formed the Chicago Board of Trade (CBOT) in 1848. The aims were to maintain a central market that treats market participants equally and fairly, to collect and disseminate commodity and economic information, and to establish quality control standards for all deliverable grades of grain. With the futures market an efficient mechanism has been established to allocate price risk in commodities among market participants.[1]

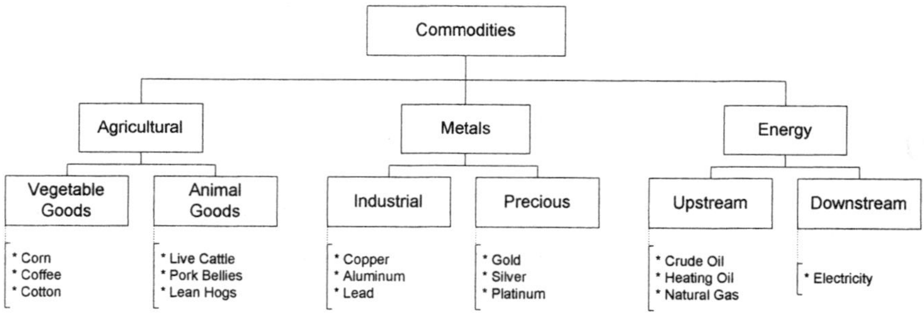

Fig. 14.1. Classification of Commodity Futures Underlyings

Today, futures exchanges exist at different places worldwide specialized in various types of futures contracts. According to their different underlyings the futures contracts can be grouped into the two major categories of *financial futures contracts* and *commodity futures contracts.*[2] In our particular study we are interested in the commodity futures markets where we further focus our attention on energy markets. In order to get an overview of commodities that are relevant in futures markets see figure 14.1 which follows the systematization of the Standard International Trade Classification.[3]

The *agricultural* futures contracts show the longest history with the major exchanges being the Chicago Board of Trade (CBOT), the Chicago Mercantile Exchange (CME), and the Coffee, Sugar, and Cocoa Exchange (CSCE). Futures contracts are traded on vegetable goods such as grains (corn, oats, wheat, and rice), foodstuffs (coffee, cocoa, orange juice, and sugar), textiles (cotton), oil and meal (soybeans, -meal, -oil, sunflower seed and oil), timber products (lumber and plywood) and on animal goods such as live hogs, live

[1] For historical, institutional and regulatory issues of futures exchanges see, for example, Markham (1987).
[2] See, for example, Duffie (1989).
[3] See UN (1996).

cattle, lean hogs, and pork bellies. For many of these commodities various futures contracts are available due to the necessary standardization of futures contracts. The specification of contracts for different grades and types of the commodity ensures a homogeneous deliverable commodity. In the *metallurgical* futures market the underlying commodities are classified by either being industrial or precious metals. The spectrum ranges from copper, aluminum, palladium, lead, zinc, nickel, and tin up to the other end with silver, platinum, and gold. Major metal futures exchanges are the Commodity Exchange (COMEX), the New York Mercantile Exchange (NYMEX), and the London Metal Exchange (LME).

Among the various commodities the *energy* market is the most recent market to be transformed by derivative securities and risk management. A major reason for this late development is considered to be the postponed deregulation of energy markets giving the spot markets ultimately the necessary liquidity and price volatility. During the 1960s and 1970s prices of *oil* as one of the most crucial commodities in the world were tightly controlled by the Organization of Petroleum Exporting Countries (OPEC) together with large globally integrated oil companies. The trading of oil futures at the NYMEX began with heating oil futures in 1978, followed by crude oil futures introduced in 1983, and unleaded gasoline futures introduced in 1985 establishing the NYMEX as the world's leading energy futures exchange. A further major energy commodity besides oil is natural *gas* which accounts for almost a quarter of total US energy consumption. The gas industry has been deregulated since the enactment of the Natural Gas Policy Act of 1978 changing it to an industry that largely operates as a free market today. An accompanying futures contract to the natural gas spot market centers was launched in 1990 traded on the NYMEX.

The transformation of the electricity market into a competitive market has been essential for the development of *electricity futures* markets. Since electricity is considered a regional commodity not only deregulation in electricity markets but also in the transmission system has been crucial in creating a electricity derivatives market. Vigorous restructuring and deregulation efforts have been undertaken by both the Federal Energy Regulatory Commission (FERC) and key state commissions.[4] The passage of FERC Order 888 in 1997, along with the passage of the Public Utility Regulatory Policies Act (PURPA) of 1978 and the Energy Policy Act (EPAct) of 1992, opened the

[4] For a detailed record of deregulation history in the US power market at the federal and the state levels see, for example, Bell and Lilyestrom (1997).

bulk power transmission system to competition and has allowed new market participants to enter the generation and wholesale power business. The state-by-state process of deregulation began in California where the Government signed restructuring legislation in 1996 with the aim of developing the first competitive marketplace for electricity in the US. The revised Californian market structure established two new entities by 1998, the California Power Exchange (CalPX) to conduct an electricity auction market and an Independent System Operator (CaISO) to manage transmission from generation sources to power marketers and finally consumers.[5] Besides the centralized dispatch and real time market pricing of the CaISO, the CalPX created the first one-hour and one-day forwards on electricity.[6] For the rapidly developing cash electricity markets in the Western US exchange traded futures contracts were launched by the NYMEX in 1996, one contract based on delivery at the California-Oregon border (COB) and another for the Palo Verde switchyard in Arizona. This early introduction of electricity futures contracts is also seen critical in that a maturely evolving cash market provides the benefit of being able to identify the most desirable trading locations and the most demanded trade terms and quantities.[7] However, with the high volatility observed in deregulated electricity prices high potential of commercial applications for electricity futures in risk management are seen: Power generating companies can sell futures contracts to lock in a specific sales prices for the electricity they intend or expect to produce in the future; by buying futures power purchasers, such as large utilities or major industrial concerns, can protect their purchase prices; power marketers who have exposure on both the generating and delivery sides of the market can hedge their risks using futures contracts. But surely the risk managing of electricity derivatives themselves is a challenge; see, for example, Kaminski (1997).

14.3 Pricing Commodity Futures

In the previous section we presented the variety of different commodities serving as underlying assets for futures contracts. Across all these commodities ranging from agricultural products to pure financial assets certain common principles of futures valuation and futures price behavior apply.[8] As we demonstrate further on, the essence of the price relation between a futures contract

[5] See CalPX (2000a).
[6] See CalPX (2000b).
[7] See Johnson and Sogomonian (1997).
[8] See, for example, Duffie (1989) and Stoll and Whaley (1993).

and its underlying spot asset is captured by an arbitrage argument; this argument, however, need to be relaxed when it comes to pricing electricity derivatives.[9] For the valuation of futures prices in an arbitrage-free world we assume deterministic risk free interest rates which makes forward prices equal to futures prices.[10]

Definition 14.1 (Futures Price) *The futures price $F(t,T)$ of an asset S_t for a futures contract with delivery at time T is given by the formula*

$$F(t,T) = \mathbb{E}^{\mathbb{Q}}\left[S_T | \mathcal{F}_t\right], \tag{14.1}$$

i.e. todays arbitrage-free futures price $F(t,T)$ is the expected spot price S_t at delivery under the martingale measure $\mathbb{Q}$.

The futures pricing formula of equation (14.1) is arbitrage-free in that there are possibilities of riskless profits if the relationship is violated. This can be directly recovered from a cash-and-carry and a reverse cash-and-carry arbitrage argument. Taking into account carry costs C^C and carry returns C^R when holding the deliverable asset the pricing relationship of equation (14.1) yields the *standard cost-of-carry formula* for futures pricing:

$$F(t,T) = S(t) + \mathbb{E}^{\mathbb{Q}}\left[\left.\int_t^T C^C(s) - C^R(s)\,ds\right| \mathcal{F}_t\right]. \tag{14.2}$$

The two arbitrage arguments are as follows: First, if equation (14.2) is violated in that the futures price is strictly larger than the expected spot price including carry costs and returns under the martingale measure $\mathbb{Q}$, investors follow a *cash-and-carry arbitrage.* The arbitraguer borrows funds to buy the spot asset, takes a short position in a futures contract, and carries the deliverable asset until the futures delivery date. Second, if the futures price is below the discounted expected spot price a *reverse cash-and-carry arbitrage* can be followed. The arbitraguer sells the spot asset short, lends the proceeds, and takes a long position in a futures contract. Together both arbitrage possibilities ensure that the equality of equation (14.2) holds. However, in view of implementing the two arbitrage arguments there is a marked difference. The

[9] See the exposition in chapter 18 for the peculiarities in the pricing of derivatives on a non-storable underlying such as electricity.

[10] For a rigorous examination of the relationship between futures and forward prices see Cox, Ingersoll, and Ross (1981b), Jarrow and Oldfield (1981), and Richard and Sundaresan (1981).

cash-and-carry arbitrage can be set up and pursued easily with financial underlyings as well as storable commodities in that it only asks for carrying the deliverable asset until the futures delivery date. Thus, the cost-of-carry model can be used to determine an upper bound on futures prices in that the futures price $F(t,T)$ can not exceed the cost of stored underlying assets for delivery. If in this case the future-spot price relationship is mainly characterized by the carry costs the futures market is said to be in *contango*. This is the standard market situation where carry costs are higher than returns from carrying the asset and the futures prices lie above the spot prices. However, in the case of reverse cash-and-carry arbitrage we need to sell the deliverable asset short in order to realize riskless profits. The implementation of this arbitrage can either be complicated or impossible when there are no or little amounts of deliverable assets available in storage.[11] Thus, the cost-of-carry model can not provide a lower bound for futures pricing. The *modified cost-of-carry formula* for non-carry commodities resolves this pricing problem by including a convenience yield C^Y:[12]

$$F(t,T) = S(t) + \mathbb{E}^{\mathbb{Q}}\left[\left.\int_t^T C^C(s) - C^R(s) - C^Y(s)\,ds\right|\mathcal{F}_t\right].$$

The convenience yield is the return that users of the commodity realizes for carrying inventory of the spot commodity. This scenario of positive returns or negative storage costs arises when the users will not be willing to sell their inventory in the spot market and buy futures contracts that will replenish their supplies at a later date. A commodity's convenience yield can be different for various users and can vary over time. The return will be at highest when there are spot shortages of the spot commodity. In instances where the convenience yield plays a major role for users the spot prices will lie above futures prices, i.e. a market situation which is known as *backwardation*.

The evolving literature on valuation models for pricing commodity futures in a continuous-time framework is summarized in table 14.1. The no-arbitrage pricing literature can broadly be categorized into the two approaches of arbitrage models and equilibrium models, which of course both result in no-arbitrage pricing relationships.

---

[11] The reverse cash-and-carry arbitrage also breaks down in cases of storable commodities for which there exist no leasing market as, for example, the natural gas market. See, for example, Leong (1997).

[12] The concept of a convenience yield was first described by Kaldor (1939) and Working (1949).

Table 14.1. Commodity Pricing Models

Authors	Model Specification	Constants
Brennan/ Schwartz (1985)	$dS = \mu S dt + \sigma S dz$ $C(S,t) = cS$	$\mu, \sigma,$ c
Gibson/ Schwartz (1990)	$dS = (\mu - \delta) S dt + \sigma_1 S dz_1$ $d\delta = \kappa(\alpha - \delta) dt + \sigma_2 dz_2$ $dz_1 dz_2 = \rho dt$	$\mu, \sigma_1,$ $\kappa, \alpha, \sigma_2,$ ρ
Brennan (1991)	$dS = \mu S dt + \sigma S dz_s$ $dC = \alpha(m - C) dt + \eta dz_c$ $dz_s dz_c = \rho dt$	μ, σ α, m, η ρ
Schöbel (1992)	$dI = \kappa\left(\ln \overline{I} - \ln I\right) I dt + \sigma I dw$ $P = \overline{P}\left(I/\overline{I}\right)^{-\varphi}$	$\kappa, \overline{I}, \sigma,$ φ
Ross (1997)	$dS = k(\theta - S) dt + \sigma(.) dz$	k, θ, σ
Schwartz (1997, M. 1)	$dS = \kappa(\mu - \ln S) S dt + \sigma S dz$	κ, μ, σ
Schwartz (1997, M. 3)	$dS = (r - \delta) S dt + \sigma_1 S dz_1^*$ $d\delta = \kappa(\widehat{\alpha} - \delta) dt + \sigma_2 dz_2^*$ $dr = (m^* - r) dt + \sigma_3 dz_3^*$ $dz_1^* dz_2^* = \rho_1 dt, dz_2^* dz_3^* = \rho_2 dt, dz_1^* dz_3^* = \rho_3 dt$	$\sigma_1,$ $\kappa, \widehat{\alpha}, \sigma_2,$ $m^*, \sigma_3,$ ρ_1, ρ_2, ρ_3
Clewlow/ Strickland (1999)	$dS = \alpha(\mu - \ln S) S dt + \sigma S dz$ $\mu = \frac{\partial \ln F(0,1)}{\partial t} + \ln F(0,1) + \frac{\sigma^2}{4}\left(1 - e^{-2\alpha t}\right)$	α, σ
Schwartz/ Smith (2000)	$d\chi_t = -\kappa \chi_t dt + \sigma_\chi dz_\chi$ $d\xi_t = \mu_\xi dt + \sigma_\xi dz_\xi$ $\ln S_t = \chi_t + \xi_t \quad dz_\chi dz_\xi = \rho_{\chi\xi} dt$	$\kappa, \sigma_\chi,$ $\mu_\xi, \sigma_\xi,$ $\rho_{\chi\xi}$

In the class of *arbitrage models* the underlying commodity is assumed to be tradable. The first stochastic futures pricing model of this class is presented by Brennan and Schwartz (1985) who use a convenience yield $C(S,t)$ that is proportional to the spot asset price. They illustrate the nature of their solution on a copper mine using stylized facts. However, the assumption of a constant convenience yield only holds under restrictive assumptions, since the theory of storage is rooted in an inverse relationship between the convenience yield and the level of inventories. The proposed models by Gibson and Schwartz (1990) and Brennan (1991) therefore extend the single-factor models to two-factor models with a mean-reverting stochastic specification of the convenience yield denoted, respectively, by δ and C. In extending these approaches Schwartz (1997, Model 3) presents a three-factor model which includes a stochastic risk free rate of interest r. He implements the pricing model on empirical data of oil futures contracts. The paper by Hilliard and Reis (1998) investigates the impact of stochastic convenience yields, stochastic interest rates, and Poisson-distributed jumps in the spot price process in the valuation of commodity futures, forwards, and futures options. Finally, Casassus and Collin-Dufresne (2002) develop a three-factor model of commodity spot prices, convenience yields, and interest rates which nests previously proposed models and allows for time-varying risk premia.

The valuation approach common to these pricing models is their reliance on the information content revealed by the spot prices and their foundation in the *theory of storage*. Instead of using price information available at the spot commodity markets there are pricing models that use the commodity futures or forward *price curve*, which is observable at the market, as given. Similar to the technique of modeling the evolution of the entire forward curve known from interest rate derivatives pricing, Clewlow and Strickland (1999) derive analytical pricing formulae for standard commodity derivatives.[13] Based on a mean-reverting specification of the commodity spot price process they develop a trinomial tree approach and are thereby able to price general path-dependent derivative contracts. The approach of modeling the movement of the term structure of commodity futures prices is also followed by Cortazar and Schwartz (1994) and Amin, Ng, and Pirrong (1995).

The category of *equilibrium models* assumes the underlying commodity to be a non-traded asset. Dealing with an underlying asset that is not traded the usual hedging argument in pricing derivative assets does not apply and the value of a futures contract is determined by the expected spot price at maturity of the contract. One-factor models based on a dynamic specification of the spot price behavior are presented by Ross (1997) and Schwartz (1997, Model 1). Ross (1997) considers an underlying asset which cannot be stored and held over time except at high cost making the pricing of derivative contracts impossible with the usual hedging argument. The quite general volatility specification of Ross (1997) is put in concrete terms by Schwartz (1997, Model 1) who assumes a volatility that is linear in the spot price. In the two-factor model of Schwartz and Smith (2000) the log-spot price behavior is decomposed into a process to capture the equilibrium level ξ_t and a dynamic specification of transitory disturbances χ_t from the long-run level.[14]

Furthermore, *regime switching models* are proposed in the literature which are hybrid approaches grasping both arbitrage and equilibrium pricing situa-

[13] This modeling approach can be traced back to Ho and Lee (1986) and Heath, Jarrow, and Morton (1992).

[14] Schwartz and Smith (2000) show that their model is equivalent to the convenience yield model of Gibson and Schwartz (1990).

tions of futures contracts. The regime switching approach of Schöbel (1992) models the spot price P via the dynamics of the level of inventories I. Dependent on the availability of the spot asset the model captures a complete and an incomplete market situation which are given for a spot asset that is, respectively, available or too scarce for arbitrage purposes. Based on Deaton and Laroque (1992) and Deaton and Laroque (1996), Routledge, Seppi, and Spatt (2000) determine spot and forward prices endogenously from an immediate net-demand process and the resulting dynamic structure of equilibrium inventory.[15] They additionally discuss a tractable two-factor augmentation of their model which leads to better calibration results of the volatility term structures. In the paper by Bühler, Korn, and Schöbel (2000) a unifying approach is presented in that conditional on the levels of current spot prices and inventories the derived futures prices span a wide price range which is equivalently obtainable from using a cost-of-carry model up to using a pure equilibrium model.

14.4 Asset Pricing in Electricity Markets

Before we develop our pricing model in the next chapter, we examine the existing literature. However, when comparing the pricing of electricity derivatives to the existing asset pricing models for traditional commodities as presented in the previous section we first need to encompass two unique characteristics of electricity: The *non-storability* of power and the *specific price patterns* of deregulated electricity prices.

Electricity as a commodity has many interesting characteristics, most of which stem from the fact that it cannot be economically stored. This peculiarity of *non-storability* of electricity has several crucial implications for pricing derivative assets:

(i) The non-storability feature means that due to a lack of any power inventories electricity must be produced at exactly the same time as it is consumed. This creates a load-matching problem in that the utility industry needs to discover the value of lost load and decide on the optimum amount of reserve margin to provide.[16]
(ii) The non-storability problem could be diminished if there would be a physical possibility of economically transferring power from an overproduction

[15] However, it is not a continuous-time pricing model but formulated in discrete time.
[16] See, for example, Woodley and Hunt (1997, p. 44 f.).

region to a consumer region. However, the US power markets are geographically distinct with several regions serving as delivery points for electricity futures contracts. Currently, there are no satisfactory transmission policies but inter-regional price differences will decrease as deregulation proceeds and the utility industry is required to open up its transmission systems.[17]

(iii) If electricity is practically non-storable, no inventories can be hold and thereby it is impossible to specify the positive returns from owning the commodity for delivery if power is not storable. Thus, the convenience yield models can hardly be extended to price electricity derivatives. This incomplete market situation asks for different pricing models to value electricity derivatives.[18]

(iv) Due to the load-matching problem electricity prices show highly volatile seasonal, weekday and even intra-day patterns. These patterns persist since there are no arbitrage possibilities due to the non-storability feature of electricity. The hourly changing price behavior needs also to be considered in specifying appropriate electricity derivative contracts which are useful for pricing, hedging, and risk management applications.[19]

(v) Finally, the fact of non-storability has an enormous impact on the investment and operating decisions of utilities and other major producers and consumers of electricity. Thus, the non-storability plays a crucial role in determining the relevance and the range of applications of electricity derivatives.[20]

All these implications need to be considered with their impact on the pricing of electricity derivatives. To get a better understanding of the specific price patterns of *deregulated electricity prices* we first take a quick look at the deregulation background of the electricity industry.[21] The ongoing deregulations of electricity is transforming a once strictly regulated and monopolistic industry with low volatility prices being a function of generation, transmission, and distribution cost into a very active market. This market is characterized by competition in generation and supply activities, but not in trans-

[17] On the isssue of rationalising the US power transmission business see, for example, Barber (1997).

[18] For suggestions of different pricing models see, for example, Eydeland and Geman (1998).

[19] For the marked volatility in energy prices that is both high and variable over time see, for example, Duffie and Gray (1995) and Pokalsky and Robinson (1997).

[20] See, for example, Pokalsky and Robinson (1997) and Leong (1997).

[21] See, for example, Wolak (1997) and Borenstein (2002).

mission and distribution since they are considered natural monopolies. The main consequence is that electricity prices in what is usually called a market pool are determined by the supply side (electricity producers) and the demand side (electricity retailers). The generators compete to sell electricity in the pool while the power retailers, i.e. the suppliers to consumers, purchase electricity from the pool at equilibrium prices. These prices are set by the intersection of aggregated supply and demand on an hourly basis. These new deregulated electricity prices are characterized by their *price patterns* being mean-reverting and extremely volatile with spikes, besides their seasonal behavior over the year. These distributional and temporal properties of electricity price distributions which need to be captured by a proper specification of an asset pricing model have been examined in the literature. Bhanot (2000) analyzes US electricity prices from twelve regional markets focusing on their seasonal and mean-reverting nature and on possible regional differences among them. The paper by Knittel and Roberts (2001) examines the electricity price process for Californian data in a non-parametric framework and analyzes the forecasting ability of several different dynamic models. They find a high degree of persistence in the electricity price level, time-varying volatility behavior, and pronounced intraday, day of the week, and seasonal cycles. An inverse leverage effect is also revealed, where positive price shocks increase price shocks more than negative shocks. Furthermore, they include findings on a model that includes weather data in form of temperature data. The important seasonal patterns are also integrated in the study of Lucia and Schwartz (2002) which reveals its partly explaining pricing power for the Scandinavian electricity market.[22] The paper by Escribano, Pena, and Villaplana (2002) analyzes the introduced electricity price uncertainty by the deregulation of the electricity sectors in Argentina, Australia, New Zealand, Scandinavia, and Spain. They present a general model that incorporates the factors of seasonality, mean-reversion, GARCH behavior, and time-dependent jumps. The international findings are that electricity prices are mean-reverting with strong volatility and show jumps of time-dependent intensity even after adjusting for seasonality. Thus there is evidence from the existing empirical literature that the most relevant *idiosyncratic patterns* of electricity prices are *seasonality*, *mean-reversion*, and *extreme volatility with spikes* which we need to keep in mind for asset pricing purposes in electricity markets.

[22] For an examination of seasonal patterns in agricultural commodities see, for example, Sorensen (2002).

The *asset pricing literature* put forward so far in this relatively new but fast growing financial valuation segment can be categorized into two approaches: The physical models and the financial models. The category of *physical models*, also known as *operations research* or *production cost type models*, include models which simulate the physical operation of the generation and transmission system of a single utility or a whole regional electricity market.[23] Using certain variables such as fuel prices, hydro condition and load the models are capable of developing different market price scenarios of power possibly with accompanying statements about probability distributions. These models seem to be more applicable for decision making in business environments and political economics rather than for financial asset pricing purposes.[24]

Table 14.2. Selected Financial Electricity Pricing Models

Authors	Model Specification	Constants
Pilipovic (1998)	$dS = \kappa(L - S)\,dt + \sigma S dz$ $dL = \mu L dt + \xi L dw$ $dzdw = 0$	$\kappa, \sigma,$ μ, ξ
Pirrong / Jermakyan (2000)	$dq = (\mu + k(\ln q - \theta))\,qdt + \sigma_q q du - dL$ $df = \alpha_f(f,t)\,f dt + \sigma_f(f,t)\,f dz$ $dudz = \rho_{qf} dt$	k, q ρ_{qf}

A second category is build by the *financial models* which derive fair values for electricity derivative prices using standard contingent claim pricing techniques based on the arbitrage pricing theory. A continuous-time valuation model for electricity forwards is presented by Pilipovic (1998) - see table 14.2 - who derives a factor pricing model based on a mean-reverting specification of the electricity spot price S and its long-run equilibrium value L. A proposed implementation of the model uses the observable electricity forward curve to calibrate the model parameters to the market volatility matrix. Eydeland and Geman (1998) present a pricing model for electricity options that relies on assumptions regarding the evolution of forward power prices. In light of the time-dependent variations in power prices and the non-linear

[23] See, for example, Jameson (1997).

[24] In applying these models for pricing purposes it needs to be ensured that power derivatives are valued risk neutrally. Leong (1997), for example, sees the need of properly distinguishing between power forward prices and forecasts of power prices.

relation between load and spot power prices Pirrong and Jermakyan (2000) develop an equilibrium pricing model where the spot price is a function of two state variables.[25] For the independent variables they choose the load q and the fuel price f resulting in an equilibrium contingent claim pricing model where they especially care about the non-storability of power. However, in contrast to, for example, Bessembinder and Lemmon (2002) they do not attempt to model the determinants of the market price of power risk. For stochastic models of energy commodity prices incorporating the possibility of jumps see, for example, Deng (2000) who includes pricing formulae for various electricity derivative assets using Fourier transformation methods.[26] The paper by Routledge, Seppi, and Spatt (2001) presents a competitive rational expectations model of spot and forward prices for multiple commodities which can be stored and/or converted. Focusing on the linkage between the markets for natural gas and electricity (the 'spark spread') their model extends the theory of storage and equilibrium models. With the conversion option they are able to consider upstream and downstream, i.e. electricity, commodities. They obtain several implications, including predicting mean-reversion of spot prices, positively skewed and heteroskedastic electricity price distributions, and unstable electricity-fuel price correlations. The authors Lucia and Schwartz (2002) propose and estimate one- and two-factor mean-reverting stochastic models in the spirit of Schwartz and Smith (2000) with deterministic seasonal cycles for the Scandinavian electricity market. Their study reveals that the seasonal pattern of spot electricity prices, which they estimate prior to derivatives pricing, plays a central role in explaining part of the shape of the term structure of futures prices. The equilibrium based forward pricing model of Bessembinder and Lemmon (2002) implies that the forward electricity price will generally be a biased forecast of the future spot electricity price. They find evidence for the forward premium decreasing by the anticipated variance of spot prices while increasing by the anticipated skewness of spot prices. The model confirms that forward prices will exceed expected spot prices when either expected demand or demand volatility are high, due to the positive skewness in the spot price distribution. For electricity forward prices the empirical studies of Longstaff and Wang (2002) and Longstaff and Wang (2003) for US prices find significant forward premia varying systematically throughout the day and that are directly related to economic risk factors. Their findings support the hypoth-

[25] For the stochastic specifications see table 14.2.

[26] For Fourier analysis techniques applied to problems of financial asset pricing see, for example, Zhu (2000).

esis that electricity forward prices are determined rationally by risk averse economic agents.

15

Electricity Pricing Model

15.1 Model Assumptions and Risk-Neutral Pricing

A proper specification of the model assumptions goes along with the decision on the pursued pricing objective. Our further attention concentrates on pricing short-term electricity derivatives, a choice which is partly motivated by the scarcely available and often illiquid data history on electricity derivatives.

Our choice is the Californian power market which provides historical spot as well as forward price data.[1] In figure 15.1 we graph the power prices of electricity in California for the year of 1999. Depicted are daily spot and day-ahead prices for 2 p.m. From the graph we note that the spot and forward prices exhibit non-linearities. The prices show extreme spikes as well as high volatility which changes rapidly over short time periods which corresponds to the findings in the empirical literature cited in section 14.4.

For the selected pricing segment of day-ahead forwards we develop a pricing model that incorporates a stochastic volatility pattern following the suggestions in the current literature:

> "Prices in the energy markets are marked by a volatility that is both high and variable over time. These characteristics mean that [...] the ability to track and forecast volatility is of paramount importance when trading and hedging energy related portfolios of derivatives."[2]
> "Stochastic volatility is certainly necessary if we want a diffusion representation to be compatible with the extreme spikes as well as the fat tails displayed by the distribution of the realized power prices."[3]

[1] See section 16.2 for a detailed description of the selected data sample and the estimation results of the empirical inferences.

[2] Duffie and Gray (1995, p. 39).

[3] Eydeland and Geman (1998, p. 73).

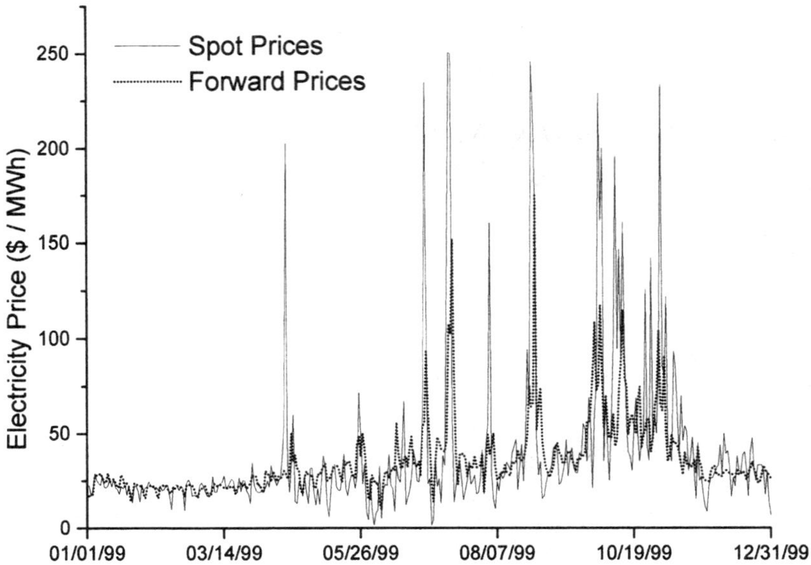

Fig. 15.1. Electricity Spot and Day-Ahead Forward Prices in California

In developing a pricing model for short-term electricity derivatives, we consider the characteristic price patterns of seasonality and mean-reversion modeled on the spot price dominated by a *stochastic specification of the spot volatility*. However, for long-term electricity contracts we find it necessary to model the spot price by a mean-reverting diffusion with seasonal patterns.[4] In the literature various stochastic volatility models are suggested, including the dynamic no-arbitrage models by Hull and White (1987), Stein and Stein (1991), and Schöbel and Zhu (1999). The stochastic volatility model we propose and examine further on is based on the in the following model assumptions.

Assumption 15.1 (State Variables) *We specify the stochastic dynamics for the state variables S_t and v_t, with S_t being the spot price of electricity and v_t the variance rate of the spot price, as in Heston (1993) by the processes under the objective probability measure $\mathbb{P}$.*

$$dS_t = \mu S_t dt + S_t \sqrt{v_t} dW^{\mathbb{P}}_{S,t}, \text{ and} \tag{15.1}$$

$$dv_t = \mu_v dt + \sigma \sqrt{v_t} dW^{\mathbb{P}}_{v,t}, \tag{15.2}$$

[4] For an analysis on the impact of mean-reversion processes on, for example, interest rate contingent claims see Uhrig-Homburg (1999).

with the drift of the variance rate process specified as $\mu_v = \kappa(\theta - v_t)$. *We further assume that* $dW^{\mathbb{P}}_{S,t} dW^{\mathbb{P}}_{v,t} = \rho dt$ *to cover possible correlation between the two state variables.*

Examining the stochastic behavior of S_t assumed in equation (15.1) shows that the spot price follows a geometric Brownian motion with a stochastic specification of the volatility term. In our case of developing a valuation model for pricing electricity forwards we need to modify the standard Heston (1993) valuation model which is originally intended to price equity, bond, and currency claims.

Assumption 15.2 (Electricity Extension) *The extension of the Heston (1993) model is based on the idea of Ross (1997) on equilibrium pricing of assets that cannot be stored and held over time except at high cost. Given electricity as an underlying commodity that is non-storable, we face a market situation that is incomplete. In this case both state variables - the spot price of electricity* S_t *and its variance rate* v_t *- cannot be hedged.*[5] *In thus, the state variables must be risk adjusted by incorporating market prices of risk in order to develop arbitrage-free pricing formulas for electricity derivatives.*[6]

Using the market prices of risk, we are able to derive the risk-neutral dynamics of the state variables in order to price electricity derivatives under the corresponding martingale measure in an incomplete market. In order to obtain the corresponding martingale process for the electricity spot price, we use the Girsanov transformation $dW^{\mathbb{Q}}_{S,t} = dW^{\mathbb{P}}_{S,t} + \lambda^* \sqrt{v_t} dt$ with a time invariant market price of risk λ^*. Based on equation (15.1) the stochastic development of the transformed spot price $X_t = \ln S_t$ leads under the Girsanov transformation to the stochastic behavior

$$dX_t = [\mu - \lambda v_t]\, dt + \sqrt{v_t} dW^{\mathbb{Q}}_{S,t} \tag{15.3}$$

for the first model factor using the substitution $\lambda = \frac{1}{2} + \lambda^*$. In the case of the spot price variance rate v_t as the second state variable, we use the same adjustment as suggested in Heston (1993) . This is given by

$$dv_t = [\kappa(\theta - v_t) - \lambda_v(S_t, v_t, t)]\, dt + \sigma \sqrt{v_t} dW^{\mathbb{Q}}_{v,t} \tag{15.4}$$

[5] In Heston (1993) the spot price diffusion models the dynamics of a stock price which is a tradeable and therefore hedgeable asset.

[6] On the relevance of incorporating risk premia in the valuation of power derivatives see, for example, Pirrong and Jermakyan (2000), Lucia and Schwartz (2002) and Bessembinder and Lemmon (2002).

with the market price of risk being $\lambda_v(S_t, v_t, t) = \lambda_v v_t$ where λ_v is a constant. Based on the risk neutralized processes (15.3) and (15.4) of the two state variables we are able to price forward contracts arbitrage-free using the EMM-approach as we demonstrate in the following section, using martingale methods developed by Geman, Karoui, and Rochet (1995) and Scott (1997).

15.2 Valuation of Electricity Forwards

Let the current market price of a forward contract with time $T - t$ until maturity be denoted by $F(t, T, S_t, v_t; \psi)$ in our model, where ψ denotes the vector of model parameters. Assuming deterministic interest rates, we know from equation (14.1) we know that forward prices are equal to the expected future spot price under the risk-neutral measure $\mathbb{Q}$

$$F(t, T, S_t, v_t; \psi) = \mathbb{E}^{\mathbb{Q}}\left[e^{X_T} | \mathcal{F}_t\right], \tag{15.5}$$

although the forward contracts for electricity cannot be priced using the standard cost-of carry arbitrage relationships due to the non-storability of electricity.

To calculate the expectation of equation (15.5) we first derive the integral solutions of the log-transformed spot state variable X_t. Based on the stochastic behavior of the X_t–process in equation (15.3) we obtain the solution

$$X_T = X_t + \mu(T - t) - \lambda \int_t^T v_t dt + \int_t^T \sqrt{v_t} dW_{S,t}^{\mathbb{Q}} \tag{15.6}$$

for the relevant time horizon $[t, T]$. Therein, we can further get an expression for the time integral from the variance rate specification of equation (15.4). Working analogously with an integral solution we get

$$\int_t^T v_t dt = \frac{1}{\kappa + \lambda_v}\left[\kappa\theta(T - t) - (v_T - v_t) + \sigma \int_t^T \sqrt{v_t} dW_{v,t}^{\mathbb{Q}}\right]. \tag{15.7}$$

Using the results of equations (15.6) and (15.7) we intermediately resolve the following expression for the expectation of equation (15.5):

$$F(t, T) = \mathbb{E}^{\mathbb{Q}}\left[e^{X_t + \mu(T-t) - \frac{\lambda}{\kappa + \lambda_v}\left[\kappa\theta(T-t) - (v_T - v_t) + \sigma\int_t^T \sqrt{v_t} dW_{v,t}^{\mathbb{Q}}\right] + \int_t^T \sqrt{v_t} dW_{S,t}^{\mathbb{Q}}} \middle| \mathcal{F}_t \right].$$

Therein, we need to take a further look at the two remaining correlated[7] stochastic integrals in the exponent. To compare these two martingales we use the fact that the transformation

$$dW_{S,t}^{\mathbb{Q}} = \rho dW_{v,t}^{\mathbb{Q}} + \sqrt{1-\rho^2} dW_t^{\mathbb{Q}}$$

results in two uncorrelated processes, i.e. $dW_{v,t}^{\mathbb{Q}} dW^{\mathbb{Q}} = 0$. This brings us to

$$F(t,T) = \mathbb{E}^{\mathbb{Q}} \left[e^{X_t + \mu(T-t) - \frac{\lambda}{\kappa+\lambda_v}\left[\kappa\theta(T-t) - (v_T - v_t) + \sigma\int_t^T \sqrt{v_t} dW_{v,t}^{\mathbb{Q}}\right]} \right. \\ \left. e^{\rho\int_t^T \sqrt{v_t} dW_{v,t}^{\mathbb{Q}} + \sqrt{1-\rho^2}\int_t^T \sqrt{v_t} dW_t^{\mathbb{Q}}} \middle| \mathcal{F}_t \right], \tag{15.8}$$

in which we can work with Itô's isometry on the last integral since the Brownian motion $W_t^{\mathbb{Q}}$ is independent of the variance rate variable. Thereby, we obtain for the last term

$$\sqrt{1-\rho^2} \int_t^T \sqrt{v_t} dW_t^{\mathbb{Q}} = \frac{1}{2}\left(1-\rho^2\right) \int_t^T v_t dt$$

under the $\mathbb{Q}$–expectation in the exponent.

However, the first type integral in equation (15.8) further remains unsolved. But from equation (15.4) we are able to obtain the expression

$$\int_t^T \sqrt{v_t} dW_{v,t}^{\mathbb{Q}} = \frac{1}{\sigma}\left[-\kappa\theta(T-t) + (v_T - v_t) + (\kappa + \lambda_v)\int_t^T v_t dt\right].$$

Putting the results together and bringing the constant terms with respect to the information available in $\mathcal{F}_t$ outside the conditional expectation we get the expression

$$F(t,T) = e^{X_t - \frac{\rho}{\sigma}v_t + \left(\mu - \frac{\kappa\theta}{\sigma}\rho\right)(T-t)} \\ \mathbb{E}^{\mathbb{Q}}\left[e^{\frac{\rho}{\sigma}v_T - \left(\lambda - \frac{\kappa+\lambda_v}{\sigma}\rho - \frac{1}{2}\left(1-\rho^2\right)\right)\int_t^T v_t dt} \middle| \mathcal{F}_t \right]. \tag{15.9}$$

With equation (15.9) we have almost derived the price of the forward contract $F(t,T)$ besides solving for the remaining expectation

[7] The correlation coefficient is assumed to be $dW_{S,t}^{\mathbb{Q}} dW_{v,t}^{\mathbb{Q}} = \rho dt$.

$$y(t, v_t) = \mathbb{E}^{\mathbb{Q}} \left[\exp\left(- \int_t^T k_1 v_s ds \right) \exp\left(k_2 v_T\right) \middle| \mathcal{F}_t \right], \tag{15.10}$$

with $k_1 = \lambda - \frac{\kappa + \lambda_v}{\sigma}\rho - \frac{1}{2}\left(1 - \rho^2\right)$, and

$k_2 = \frac{\rho}{\sigma}$.

A common tool in solving such types of conditional expectations is applying the Feynman-Kac formula which states the following.

Theorem 15.1 (Feynman-Kac). *Let $\{X_t\}_{t\geq 0}$ be an Itô process, $f(X_t)$ a non-negative function and assume that q is lower bounded. Then consider the function*

$$y(t, x) = \mathbb{E}_x \left[e^{-\int_0^t q(X_s)ds} f(X_t) \right].$$

For this function the backward differential equation

$$\frac{\partial y}{\partial t} = \mathcal{A}y - q(X_t)\, y$$

holds with the necessary initial condition being $y(0, x) = f(x)$; $\mathcal{A}$ denotes the differential operator.[8]

In our case of equation (15.10) the corresponding partial differential equation under the equivalent martingale measure $\mathbb{Q}$ takes the form

$$\frac{1}{2}\sigma^2 v_t \frac{\partial^2 y(t, v_t)}{\partial v_t^2} + \left(\kappa\theta - (\kappa + \lambda_v)\, v_t\right) \frac{\partial y(t, v_t)}{\partial v_t} + \frac{\partial y(t, v_t)}{\partial t} = k_1 v_t y(t, v_t), \tag{15.11}$$

with the boundary condition for the value at maturity $y(T, v_t) = \exp(k_1 v_T)$. For the solution for the differential equation we propose the exponential affine structure

[8] The formula is named after the contributions of Feynman (1948) and Kac (1949). For a complete definition and the relevant conditions see, for example, Duffie (2001), Karatzas and Shreve (1991), and Oksendal (2000) who also describes the differential operator.

$$y(t, v_t) = \exp(A(t,T)\, v_t + B(t,T)), \tag{15.12}$$

$$\begin{aligned}
\text{with } \frac{\partial y(t, v_t)}{\partial t} &= \left(\frac{\partial A(t,T)}{\partial t} v_t + \frac{\partial B(t,T)}{\partial t}\right) y(t, v_t), \\
\frac{\partial y(t, v_t)}{\partial v_t} &= A(t,T)\, y(t, v_t), \text{ and} \\
\frac{\partial^2 y(t, v_t)}{\partial {v_t}^2} &= A(t,T)^2\, y(t, v_t),
\end{aligned}$$

including the transformed boundary conditions $A(T,T) = k_2$ and $B(T,T) = 0$.

The result of bringing equations (15.11) and (15.12) together is given by the system of two partial differential equations:

$$\frac{\partial A(t,T)}{\partial t} + \frac{1}{2}\sigma^2 A(t,T)^2 - (\kappa + \lambda_v)\, A(t,T) = k_1 \tag{15.13}$$

$$\kappa\theta A(t,T) + \frac{\partial B(t,T)}{\partial t} = 0, \tag{15.14}$$

where we applied the separation of variables technique.[9] The first equation (15.13) is known as a Riccati type differential equation[10] for which we get the solution

$$\begin{aligned}
A(t,T) &= \frac{1}{\sigma^2}\left[(\kappa + \lambda_v) - \gamma_1 \frac{\gamma_1 \sinh \frac{1}{2}\gamma_1 (T-t) + \gamma_2 \cosh \frac{1}{2}\gamma_1 (T-t)}{\gamma_1 \cosh \frac{1}{2}\gamma_1 (T-t) + \gamma_2 \sinh \frac{1}{2}\gamma_1 (T-t)}\right], \\
\text{with } \gamma_1 &= \sqrt{(\kappa + \lambda_v)^2 + 2\sigma^2 k_1}, \text{ and} \\
\gamma_2 &= \kappa + \lambda_v - \sigma^2 k_2
\end{aligned} \tag{15.15}$$

incorporating the boundary condition $A(T,T) = k_1$.[11] Given the solution to the function $A(t,T)$ of equation (15.12) we can integrate the second differential equation (15.14) as

$$B(T,T) - B(t,T) = -\kappa\theta \int_t^T A(t,T)\, dt,$$
$$\text{with } B(T,T) = 0$$

[9] See, for example, Ingersoll (1987, p. 397) for an application with interest rate contingent claims.

[10] This type of differential equation commonly arises with the squared Gaussian model of Cox, Ingersoll, and Ross (1985b); see, for example, Rogers (1995, p. 99 f.).

[11] For a similar solution to equation (15.13) see, for example, Rogers (1995, p. 100) and Stein and Stein (1991, p. 730).

to get a result for the function $B(t,T)$. Working out the integral we obtain

$$B(t,T) = \frac{\kappa\theta}{\sigma^2}\left[(\kappa+\lambda_v)(T-t) \right.$$
$$\left. -2\ln\left\{\cosh\frac{1}{2}\gamma_1(T-t) + \frac{\gamma_2}{\gamma_1}\sinh\frac{1}{2}\gamma_1(T-t)\right\}\right] \quad (15.16)$$

for the second function in equation (15.12). Thus, we have finally calculated an expression for the remaining expectation $y(t, v_t)$ of equation (15.9).

Solution 15.2 (Forward Price) *Having gone through the calculations we are now able to state our final result for the time t value of a forward contract on electricity:*

$$F(t,T,S_t,v_t;\psi) = S_t e^{\left(A(t,T)-\frac{\rho}{\sigma}\right)v_t + B(t,T) + \left(\mu - \frac{\kappa\theta}{\sigma}\rho\right)(T-t)}, \quad (15.17)$$

where the functions $A(t,T)$ and $B(t,T)$ are given by the expressions of equations (15.15) and (15.16), respectively, with $\psi = \{\mu,\kappa,\theta,\sigma,\rho,\lambda,\lambda_v\}$.

The full curve of forward prices at a given date t is given by applying different times to maturity $T-t$ to equation (15.17) for the same values of the state variables X_t and v_t.

16

Empirical Inference

16.1 Estimation Model

The pricing model we propose for the valuation of short-term electricity forwards is based on a factor specification of the spot electricity price and its variance yielding a closed-form solution for the forward curve as given in equation (15.17). To estimate such dynamic commodity pricing models, several econometric procedures have been put forward of which the Kalman filter is one of the most successful procedures. The widely used estimation methodology can handle multi-factor models with non-observable state variables and measurement errors. The Kalman filter has been employed to estimate commodity asset pricing models by Schwartz (1997) in his analysis of mean reversion in commercial commodity prices with a comparison of one-, two-, and three-factor models. Using Kalman filtering and 127 years of data Pindyck (1999) examines the long-run evolution of oil, coal and natural gas prices and addresses the questions on modeling their stochastic processes and forecasting them with models of reversion to stochastically fluctuating trend lines. The paper by Karesen and Husby (2002) presents a joint state-space model for electricity spot and futures prices. Within the general Heath, Jarrow, and Morton (1992) no-arbitrage framework and in light of Cortazar and Schwartz (1994), Amin, Ng, and Pirrong (1995), and Miltersen and Schwartz (1998) the work by Manoliu and Tompaidis (2002) presents a class of multi-factor stochastic models for energy futures prices where their application is based on Kalman filtering, maximum likelihood estimation, and futures prices for natural gas. Kalman filtering is employed in a seasonal price model of agricultural commodity futures by Sorensen (2002), which is similar to the suggested model of Schwartz and Smith (2000) but includes a deterministic seasonal component using a parametrized linear combination of trigonometric functions with seasonal frequencies. The work by Cortazar and Naranjo (2003) generalizes the existing three-factor models of Schwartz (1997) and Cortazar and Schwartz (2003) to an n-factor setting and shows estimation results on a Kalman filter framework that may be applied in an incomplete panel data setting, optimally employing all market prices available. Based on these estimation ideas we conveniently transform our proposed theoretical pricing model from chapter 15 into an equivalent empirical state space model whereupon we are able to build a Kalman filtering algorithm to calibrate our forward pricing model to available empirical data. From chapter 2 we know that a state space model consists of two major pillars: the transition and the measurement equation. In the case of our forward pricing model the transition equation captures the dynamics of the state variables. The measurement equation provides us with a link of the spot electricity price to the forward prices of electricity. In the following we first focus on specifying the transition equation by examining the distribution of the state variables.

16.1.1 Distribution of the State Variables

The distribution of the state variables is approximated by the first two exact conditional moments for the state variables. These moments are implied by the stochastic process specifications of the factors X_t and v_t. For convenience we restate the dynamic specifications of the state variables in integral form as follows:

$$\int_t^s dX_\varsigma = \int_t^s \left[\mu - \frac{1}{2}v_\varsigma\right] d\varsigma + \int_t^s \sqrt{v_\varsigma} dW_{X,\varsigma}^{\mathbb{P}}, \text{ and} \tag{16.1}$$

$$\int_t^s dv_\varsigma = \int_t^s \kappa\left(\theta - v_\varsigma\right) d\varsigma + \int_t^s \sigma\sqrt{v_\varsigma} dW_{v,\varsigma}^{\mathbb{P}}, \tag{16.2}$$

for a period of time $\tau = [t, s]$. These equations fully contain the information to derive the conditional moments of the state variables. In order to calculate the moments we employ a technique of putting up differential equations that contain the conditional moments of interest as functions. With the relevant boundary conditions we then solve these *differential equations for the moments.*[1]

[1] This method is, for example, used by De Jong and Santa-Clara (1999) in the context of a factor based interest rate model à la Heath, Jarrow, and Morton (1992).

For the *conditional first moments* we take the expectations on equations (16.1) and (16.2). To calculate the conditional mean of v_s we take the derivative with respect to the upper bound s of the integrals which yields

$$\frac{\partial \mathbb{E}^{\mathbb{P}}\left[v_s \mid \mathcal{F}_t\right]}{\partial s} = \kappa\theta + \kappa \mathbb{E}^{\mathbb{P}}\left[v_s \mid \mathcal{F}_t\right]. \tag{16.3}$$

Exploiting the initial condition of $\mathbb{E}^{\mathbb{P}}\left[v_t \mid \mathcal{F}_t\right] = v_t$ we solve the differential equation (16.3) as

$$\mathbb{E}^{\mathbb{P}}\left[v_s \mid \mathcal{F}_t\right] = \theta\left(1 - e^{-\kappa(s-t)}\right) + v_t e^{-\kappa(s-t)}, \tag{16.4}$$

which resembles the solution to the conditional expectation in the interest rate model of Cox, Ingersoll, and Ross (1985b).[2] The conditional mean of the first state variable X_t can then be derived from equation (16.1) where we take expectations:

$$\mathbb{E}^{\mathbb{P}}\left[X_s \mid \mathcal{F}_t\right] = X_t + \mu\left(s-t\right) - \frac{1}{2}\int_t^s \mathbb{E}^{\mathbb{P}}\left[v_\zeta \mid \mathcal{F}_t\right] d\zeta.$$

Using the expression for $\mathbb{E}^{\mathbb{P}}\left[v_s \mid \mathcal{F}_t\right]$ from equation (16.4) we can calculate the solution

$$\begin{aligned}\mathbb{E}^{\mathbb{P}}\left[X_s \mid \mathcal{F}_t\right] = {} & \frac{\theta}{2\kappa}\left(1 - e^{-\kappa(s-t)}\right) + \left(\mu - \frac{1}{2}\theta\right)(s-t) \\ & + X_t - \frac{1}{2\kappa}\left(1 - e^{-\kappa(s-t)}\right) v_t \end{aligned} \tag{16.5}$$

for the conditional expectation of the log-spot price X_t.

In order to derive the *second moments* of the state variables we work on the basis of the statistical relationship:

$$\begin{bmatrix} \mathbb{V}ar^{\mathbb{P}}\left[X_s^2 \mid \mathcal{F}_t\right] \\ \mathbb{C}ov^{\mathbb{P}}\left[X_s v_s \mid \mathcal{F}_t\right] \\ \mathbb{V}ar^{\mathbb{P}}\left[v_s^2 \mid \mathcal{F}_t\right] \end{bmatrix} = \begin{bmatrix} \mathbb{E}^{\mathbb{P}}\left[X_s^2 \mid \mathcal{F}_t\right] \\ \mathbb{E}^{\mathbb{P}}\left[X_s v_s \mid \mathcal{F}_t\right] \\ \mathbb{E}^{\mathbb{P}}\left[v_s^2 \mid \mathcal{F}_t\right] \end{bmatrix} - \begin{bmatrix} \mathbb{E}^{\mathbb{P}}\left[X_s \mid \mathcal{F}_t\right]^2 \\ \mathbb{E}^{\mathbb{P}}\left[X_s \mid \mathcal{F}_t\right]\mathbb{E}^{\mathbb{P}}\left[v_s \mid \mathcal{F}_t\right] \\ \mathbb{E}^{\mathbb{P}}\left[v_s \mid \mathcal{F}_t\right]^2 \end{bmatrix}. \tag{16.6}$$

Given the conditional means $\mathbb{E}^{\mathbb{P}}\left[X_s \mid \mathcal{F}_t\right]$ and $\mathbb{E}^{\mathbb{P}}\left[v_s \mid \mathcal{F}_t\right]$ we further need to calculate the first terms in equation (16.6) to get solutions for the three conditional second moments. Therefore, we first obtain the stochastic dynamics for the functions $f_1 = X_t^2$, $f_2 = X_t v_t$, and $f_3 = v_t^2$ using Itô's formula. This yields the following dynamics:

[2] Compare Cox, Ingersoll, and Ross (1985b, p. 392).

$$\begin{aligned} dX_t^2 &= (2\mu X_t + v_t - X_t v_t)\, dt + 2X_t\sqrt{v_t} dW_{X,t}^{\mathbb{P}}, \\ dX_t v_t &= \left(\kappa\theta X_t + (\mu + \rho\sigma)\, v_t - \frac{1}{2} v_t^2 - \kappa X_t v_t\right) dt \\ &\quad + v_t^{3/2} dW_{X,t}^{\mathbb{P}} + \sigma X_t \sqrt{v_t} dW_{v,t}^{\mathbb{P}}, \text{ and} \\ dv_t^2 &= \left(\left(2\kappa\theta + \sigma^2\right) v_t - 2\kappa v_t^2\right) dt + \sigma\sqrt{v_t} dW_{v,t}^{\mathbb{P}}. \end{aligned}$$

Based on these stochastic differential equations we can again - as with the conditional first moments - take expectations and differentiate with respect to s. This directly transfers these equations to a system of three differential equations with the functions being the second moments $\mathbb{E}^{\mathbb{P}}\left[X_s^2 \middle| \mathcal{F}_t\right]$, $\mathbb{E}^{\mathbb{P}}\left[X_s v_s \middle| \mathcal{F}_t\right]$, and $\mathbb{E}^{\mathbb{P}}\left[v_s^2 \middle| \mathcal{F}_t\right]$:

$$\begin{aligned} \begin{bmatrix} \frac{\partial \mathbb{E}^{\mathbb{P}}\left[X_s^2 \middle| \mathcal{F}_t\right]}{\partial s} \\ \frac{\partial \mathbb{E}^{\mathbb{P}}\left[X_s v_s \middle| \mathcal{F}_t\right]}{\partial s} \\ \frac{\partial \mathbb{E}^{\mathbb{P}}\left[v_s^2 \middle| \mathcal{F}_t\right]}{\partial s} \end{bmatrix} &= \begin{bmatrix} 2\mu & 1 \\ \kappa\theta & \mu + \rho\sigma \\ 0 & 2\kappa\theta + \sigma^2 \end{bmatrix} \begin{bmatrix} \mathbb{E}^{\mathbb{P}}\left[X_s \middle| \mathcal{F}_t\right] \\ \mathbb{E}^{\mathbb{P}}\left[v_s \middle| \mathcal{F}_t\right] \end{bmatrix} \\ &\quad + \begin{bmatrix} 0 & -1 & 0 \\ 0 & -\kappa & -\frac{1}{2} \\ 0 & 0 & -2\kappa \end{bmatrix} \begin{bmatrix} \mathbb{E}^{\mathbb{P}}\left[X_s^2 \middle| \mathcal{F}_t\right] \\ \mathbb{E}^{\mathbb{P}}\left[X_s v_s \middle| \mathcal{F}_t\right] \\ \mathbb{E}^{\mathbb{P}}\left[v_s^2 \middle| \mathcal{F}_t\right] \end{bmatrix}. \end{aligned} \tag{16.7}$$

Additionally, the corresponding initial conditions are given by $\mathbb{E}^{\mathbb{P}}\left[X_t^2 \middle| \mathcal{F}_t\right] = X_t^2$, $\mathbb{E}^{\mathbb{P}}\left[X_t v_t \middle| \mathcal{F}_t\right] = X_t v_t$, and $\mathbb{E}^{\mathbb{P}}\left[v_t^2 \middle| \mathcal{F}_t\right] = v_t^2$. Further, we solve the system of equations given in (16.7) beginning with the last equation and ending with solving the first equation. Finally, we use these results along with the statistical relationship stated in equation (16.6) to derive the following expressions for the second moments. First, we obtain the solution to the conditional variance of the second state variable from the third equation of (16.7) as:

$$\mathbb{V}ar^{\mathbb{P}}\left[v_s \middle| \mathcal{F}_t\right] = \theta\frac{\sigma^2}{2\kappa}\left(1 - e^{-\kappa(s-t)}\right)^2 + v_t\frac{\sigma^2}{\kappa}\left(e^{-\kappa(s-t)} - e^{-2\kappa(s-t)}\right), \tag{16.8}$$

which matches the formula stated in Cox, Ingersoll, and Ross (1985b, eq. 19). Given the result of equation (16.8) we are further able to calculate the covariance term of the state variables as

$$\begin{aligned}
\mathbb{C}ov^{\mathbb{P}}\left[X_s v_s \middle| \mathcal{F}_t\right] = \theta &\left[\frac{\rho\sigma}{\kappa}\left(1-e^{-\kappa(s-t)}\right) - \left(\rho\sigma - \frac{\sigma^2}{2\kappa}\right)(s-t)\, e^{-\kappa(s-t)}\right. \\
&\left. -\frac{\sigma^2}{4\kappa^2}\left(1-e^{-2\kappa(s-t)}\right)\right] \\
&+v_t\left[\left(\rho\sigma - \frac{\sigma^2}{2\kappa}\right)(s-t)\, e^{-\kappa(s-t)}\right. \\
&\left. +\frac{\sigma^2}{2\kappa^2}\left(e^{-\kappa(s-t)} - e^{-2\kappa(s-t)}\right)\right].
\end{aligned} \tag{16.9}$$

Finally, using the covariance term we obtain

$$\begin{aligned}
\mathbb{V}ar^{\mathbb{P}}\left[X_s \middle| \mathcal{F}_t\right] = \theta &\left[-\frac{1}{\kappa} + \frac{2\rho\sigma}{\kappa^2} - \frac{5\sigma^2}{8\kappa^3} + \left(1 - \frac{\rho\sigma}{\kappa} - \frac{\sigma^2}{4\kappa^2}\right)(s-t)\right. \\
&+\left(\frac{1}{\kappa} - \frac{2\rho\sigma}{\kappa^2} + \frac{\sigma^2}{2\kappa^3}\right) e^{-\kappa(s-t)} \\
&\left.+\left(-\frac{\rho\sigma}{\kappa} + \frac{\sigma^2}{2\kappa^2}\right)(s-t)\, e^{-\kappa(s-t)} + \frac{\sigma^2}{8\kappa^3} e^{-2\kappa(s-t)}\right] \\
&+v_t\left[\frac{1}{\kappa} - \frac{\rho\sigma}{\kappa^2} + \frac{\sigma^2}{4\kappa^3} + \left(-\frac{1}{\kappa} + \frac{\rho\sigma}{\kappa^2}\right) e^{-\kappa(s-t)}\right. \\
&\left. -\frac{\sigma^2}{4\kappa^3} e^{-2\kappa(s-t)} + \left(\frac{\rho\sigma}{\kappa} - \frac{\sigma^2}{2\kappa^2}\right)(s-t)\, e^{-\kappa(s-t)}\right]
\end{aligned} \tag{16.10}$$

for the conditional variance of the log-spot price X_t.

To summarize the derived results, we obtained formulas for the exact theoretical first two conditional moments of the state variables. The solutions for the conditional expectations are given in equations (16.4) and (16.5) and for the variance terms with equations (16.8), (16.9), and (16.10).

16.1.2 State Space Formulation and Kalman Filter Setup

In section 2.2.4 we introduced the concept of extended Kalman filtering in addition to the linear Kalman filter, where the transition and measurement equations are linear in the state variables and the error terms are assumed to be normally distributed. In the case of implementing our forward pricing model we choose to implement an extended Kalman filter algorithm, since the state variables $\boldsymbol{\xi}_t = [X_t, v_t]'$ do not follow a Gaussian distribution. With the hypothesized forward pricing model we can impose high theoretical modeling restrictions both in the time-series as well as in the cross-section properties. First, assuming the stochastic dynamics of the state variables following the process specifications of equations (15.1) and (15.2) we can infer the corresponding transition equation. The time-series properties of the model can be deduced from the results of the first and second moments as derived in the previous section. Second, given the explicit forward pricing formula of equation (15.17) we can implement cross-sectional restrictions on the data. There, we assume that the theoretical forward prices match the observable prices on the power exchange only with small discrepancies.

Let us begin with the general specification of the *transition equation* by

$$\boldsymbol{\xi}_t = \mathbf{h}_t\left(\boldsymbol{\xi}_{t-\Delta t}, \boldsymbol{\eta}_t\left(\boldsymbol{\xi}_{t-\Delta t}, \boldsymbol{\psi}\right), \boldsymbol{\psi}\right)$$

as formulated in section 2.2.4. The conditional expectations of equations (16.4) and (16.5) directly lead to the expression

$$\begin{aligned} \mathbf{h}_t\left(\boldsymbol{\xi}_{t-\Delta t}, \boldsymbol{\eta}_t\left(\boldsymbol{\xi}_{t-\Delta t}, \boldsymbol{\psi}\right), \boldsymbol{\psi}\right) = & \begin{bmatrix} \left(\mu - \frac{1}{2}\theta\right)\Delta t + \frac{\theta}{2\kappa}\left(1 - e^{-\kappa\Delta t}\right) \\ \theta\left(1 - e^{-\kappa\Delta t}\right) \end{bmatrix} \\ & + \begin{bmatrix} 1 & -\frac{1}{2\kappa}\left(1 - e^{-\kappa\Delta t}\right) \\ 0 & e^{-\kappa\Delta t} \end{bmatrix} \boldsymbol{\xi}_{t-\Delta t} \\ & + \boldsymbol{\eta}_t\left(\boldsymbol{\xi}_{t-\Delta t}, \boldsymbol{\psi}\right) \end{aligned} \tag{16.11}$$

for the transition equation. However, within this equation the error term $\boldsymbol{\eta}_t\left(\boldsymbol{\xi}_{t-\Delta t}, \boldsymbol{\psi}\right)$ still contains the state variable $\boldsymbol{\xi}_{t-\Delta t}$ which is not observable. Thus, we approximate the transition equation around $\left(\boldsymbol{\xi}_{t-\Delta t}, \boldsymbol{\eta}_t\right) = \left(\boldsymbol{\xi}_{t-\Delta t|t-\Delta t}, \mathbf{0}\right)$ which results in:

$$\boldsymbol{\xi}_t \approx \mathbf{h}_t\left(\boldsymbol{\xi}_{t-\Delta t|t-\Delta t}, \mathbf{0}, \boldsymbol{\psi}\right) + \boldsymbol{\Phi}_{t|t-\Delta t}\left(\boldsymbol{\xi}_{t-\Delta t} - \boldsymbol{\xi}_{t-\Delta t|t-\Delta t}\right) + \mathbf{S}_{t|t-\Delta t}\boldsymbol{\eta}_t.$$

Therein, the transition matrix is given by

$$\begin{aligned} \boldsymbol{\Phi}_{t|t-\Delta t} &= \left.\frac{\partial \mathbf{h}_t\left(\boldsymbol{\xi}_{t-\Delta t}, \boldsymbol{\eta}_t, \boldsymbol{\psi}\right)}{\partial \boldsymbol{\xi}'_{t-\Delta t}}\right|_{\left(\boldsymbol{\xi}_{t-\Delta t}, \boldsymbol{\eta}_t, \boldsymbol{\psi}\right) = \left(\boldsymbol{\xi}_{t-\Delta t|t-\Delta t}, \mathbf{0}, \boldsymbol{\psi}\right)} \\ &= \begin{bmatrix} 1 & -\frac{1}{2\kappa}\left(1 - e^{-\kappa\Delta t}\right) \\ 0 & e^{-\kappa\Delta t} \end{bmatrix}, \end{aligned} \tag{16.12}$$

and the error term matrix $\mathbf{S}_{t|t-\Delta t}$ can be derived as

$$\begin{aligned} \mathbf{S}_{t|t-\Delta t} &= \left.\frac{\partial \mathbf{h}_t\left(\boldsymbol{\xi}_{t-\Delta t}, \boldsymbol{\eta}_t, \boldsymbol{\psi}\right)}{\partial \boldsymbol{\eta}'_t}\right|_{\left(\boldsymbol{\xi}_{t-\Delta t}, \boldsymbol{\eta}_t, \boldsymbol{\psi}\right) = \left(\boldsymbol{\xi}_{t-\Delta t|t-\Delta t}, \mathbf{0}, \boldsymbol{\psi}\right)} \\ &= \begin{bmatrix} \mathbb{V}ar^{\mathbb{P}}\left[X_t \middle| \mathcal{F}_{t-\Delta t}\right] & \mathbb{C}ov^{\mathbb{P}}\left[X_t v_t \middle| \mathcal{F}_{t-\Delta t}\right] \\ \mathbb{C}ov^{\mathbb{P}}\left[X_t v_t \middle| \mathcal{F}_{t-\Delta t}\right] & \mathbb{V}ar^{\mathbb{P}}\left[v_t \middle| \mathcal{F}_{t-\Delta t}\right] \end{bmatrix} \end{aligned} \tag{16.13}$$

where we use the expressions of equations (16.10), (16.9), and (16.8) and substitute values $v_{t-\Delta t|t-\Delta t}$ for $v_{t-\Delta t}$. Thus, we can finally state our transition equation of the state space model as

$$\boldsymbol{\xi}_t \approx \mathbf{c}_t + \Phi_{t|t-\Delta t}\boldsymbol{\xi}_{t-\Delta t} + \mathbf{S}_{t|t-\Delta t}\boldsymbol{\eta}_t, \qquad (16.14)$$
$$\text{with } \mathbf{c}_t = \mathbf{h}_t\left(\boldsymbol{\xi}_{t-\Delta t|t-\Delta t}, \mathbf{0}, \boldsymbol{\psi}\right) - \Phi_{t|t-\Delta t}\boldsymbol{\xi}_{t-\Delta t|t-\Delta t}$$

where the formulae of $\mathbf{h}_t\left(\boldsymbol{\xi}_{t-\Delta t|t-\Delta t}, \mathbf{0}, \boldsymbol{\psi}\right)$, $\Phi_{t|t-\Delta t}$, and $\mathbf{S}_{t|t-\Delta t}$ are given in equations (16.11), (16.12), and (16.13).

Next, we work out the specification of the *measurement equation* where we use the forward pricing formula derived in section 15.2. The idea is to comprehensively use market information contained in both the spot electricity prices as well as the information from the forward prices of electricity. Bringing both market informations together, we formulate the measurement equation as an identity between the observable market prices and the theoretical prices. However, the two pairs of prices do not need to fully match in that we include a measurement error to allow for some discrepancies between the market and model prices. We obtain the measurement equation by exploiting the functional relationship of the state variables $\boldsymbol{\xi}_t$ with the forward price from the original pricing formula given in equation (15.17)

$$\ln F(t,T) = \left(\mu - \frac{\kappa\theta}{\sigma}\rho\right)(T-t) + B(t,T) + \left[1, A(t,T) - \frac{\rho}{\sigma}\right]\boldsymbol{\xi}_t,$$

restated in logarithmic form. Denoting the observable prices on the electricity spot and forward markets by $\mathbf{y}_t = [X_t, \ln F(t,T)]'$ we are able to use the general specification of the measurement equation

$$\mathbf{y}_t = \mathbf{a}_t(\boldsymbol{\psi}) + \mathbf{B}_t(\boldsymbol{\psi})\boldsymbol{\xi}_t + \boldsymbol{\varepsilon}_t(\boldsymbol{\psi}) \qquad (16.15)$$

as known from chapter 3. The measurement equation (16.15) is further particularized by the system matrices

$$\mathbf{a}_t(\boldsymbol{\psi}) = \begin{bmatrix} 0 \\ \left(\mu - \frac{\kappa\theta}{\sigma}\rho\right)(T-t) + B(t,T) \end{bmatrix}, \text{ and}$$
$$\mathbf{B}_t(\boldsymbol{\psi}) = \begin{bmatrix} 1 & 0 \\ 1 & A(t,T) - \frac{\rho}{\sigma} \end{bmatrix}.$$

For the moment specification of the error term $\boldsymbol{\varepsilon}_t(\boldsymbol{\psi})$ in the measurement equation (16.15) we assume that the errors have zero mean, i.e. $\mathbb{E}\left[\varepsilon_t|\mathcal{F}_{t-\Delta t}\right]=\mathbf{0}$, and are serially uncorrelated. The errors' time invariant covariance matrix is further modeled by

$$\mathbb{V}ar\left[\varepsilon_t|\mathcal{F}_{t-\Delta t}\right]=\begin{bmatrix}\sigma^2_{\varepsilon,S} & 0\\ 0 & \sigma^2_{\varepsilon,F}\end{bmatrix}.$$

This covariance specification models the noise observable in the spot market separately from the noise seen in the forward markets. Thereby, we add two more parameters, $\sigma_{\varepsilon,S}$ and $\sigma_{\varepsilon,F}$, to our parameter space ψ of the theoretical futures pricing model.

Thereby, we set up the specific characteristics of the *state space model* for the electricity spot and forward market using the transition and measurement equations from (16.14) and (16.15). Based on this state space formulation we are able to run an extended Kalman filter algorithm in order to estimate our parameter values $\psi=\{\mu,\kappa,\theta,\sigma,\rho,\lambda,\lambda_v,\sigma_{\varepsilon,S},\sigma_{\varepsilon,F}\}$ by means of maximum likelihood according to

$$\hat{\psi}=\underset{\psi\in\Psi}{\arg\max}L\left(\mathbf{y}_T,\mathbf{y}_{T-1},\dots,\mathbf{y}_1;\psi\right). \tag{16.16}$$

In our case, where the conditional normal distribution is approximated by the derived second moments of equations (16.8), (16.9), and (16.10), the log-likelihood can be used to yield quasi maximum likelihood parameter estimates.[3] The asymptotic statistical properties of quasi maximum likelihood estimates for heteroskedastic models, such as the Cox, Ingersoll, and Ross (1985b) and our stochastic volatility model, are discussed in Bollerslev and Wooldridge (1992). They show that the quasi maximum likelihood estimates are consistent and asymptotically normal under fairly weak regularity conditions. Even though the assumption of normality is violated when maximizing the normal log-likelihood, these results are obtained since the score of the normal log-likelihood has the martingale difference property when the first two conditional moments are correctly specified. Thereby, one can correctly estimate not only the mean parameters but also the second moments.

For our model the conditional means and variances should probably be different from the true moments of the system since we use a linear projection of the state variables. This linear approximation is needed to incorporate the non-normal nature of the model into the typical extended Kalman filter framework. The approximation, however, can be expected to work well because it is linearly optimal. The Monte Carlo studies in Bollerslev and Wooldridge

[3] See, for example, Gourieroux, Monfort, Renault, and Trognon (1984).

(1992) and Duan and Simonato (1999) indicate that the asymptotic results carry over to finite samples, i.e. it is reasonable to expect the parameter vector $\hat{\psi}$ to be approximately consistent and asymptotically normal. Further, the small sample properties of the quasi maximum likelihood estimates in the case of single and multi-factor models à la Cox, Ingersoll, and Ross (1985b) are studied in detail by Ball and Torous (1996), Duan and Simonato (1999), and Chen and Scott (2003). The various simulation evidence therein shows that the sampling properties of the quasi maximum likelihood estimator based on the extended Kalman filter provide satisfactory results.

In the following section we perform a general analysis of the electricity data set and present the results on empirical inferences of a one-factor pricing model and our suggested stochastic volatility valuation model.

16.2 Data Analysis and Estimation Results

In this section we apply the electricity forward pricing models to study the deregulated Californian electricity market. This first and largest totally open electricity wholesale and retail market has two core institutions, the *California Power Exchange* (CalPX) and the *California Independent System Operator* (CalSO). We briefly describe their operations on the Californian electricity market, before we come to the data analysis and present our estimation results.[4]

The *CalPX* is a non-profit, public benefit corporation which is regulated by the FERC. As a power commodity exchange it provides an efficient, competitive marketplace by conducting an open, non-discriminatory trading process for qualified electricity suppliers and purchasers. The CalPX determines the price of electricity in the hour-ahead and day-ahead markets of forward contracts to deliver a given quantity of electricity over a one hour period. For each delivery hour the exchange collects demand and generation bids in prices and quantities to settle for unconstrained market clearing prices. Since its opening on March 31, 1998 the CalPX's day-ahead market has traded between 80 and 90 percent of California's electricity market. This day-ahead market is open from 6 a.m. to 1 p.m. when sellers and buyers of electricity can submit their portfolio supply and demand bids for 24 hourly auction periods for electricity delivery on the next day. Thereupon, the market clearing prices are determined separately for each auction hour. Besides providing marketplaces

[4] The Californian power market is explained in detail in, for example, CalPX (1999).

for electricity forwards, the CalPX serves as scheduling coordinator in that it submits balanced supply and demand schedules along with information on the generating units and the locations for delivery to the CaISO.

The *CaISO* serves as the control area operator for the Californian electricity market where it provides open access to the transmission and manages the real time operations of the transmission grid. The objective is to provide reliable system operations, primarily keeping the right frequency and providing sufficient generation, which are maintained by buying and providing ancillary services. The CaISO matches the power output of the generating units with the power demand within the controlled electric power system including the scheduled interchange with adjacent control areas. This real time balancing of load and generation is based on the coordination of the hour-ahead and day-ahead schedules dispatched by the CalPX. On the basis of the real time operations, the CaISO determines and publishes real time spot electricity prices by hour.

In tables 16.1 and 16.2 we present descriptive statistics on the Californian electricity prices from the spot and day-ahead forward markets. The data are sampled separately for each hour of the day, as for example in Longstaff and Wang (2002), from the opening of the CalPX on 04/01/1998 until 12/31/1999. The *spot prices* for delivery of one MWh of electricity average around \$29.62 with a mean standard deviation of \$27.83 which is very high. The prices take a wide spectrum from prices close to zero up to values of \$492.20. Comparing the spot price averages of the mean with the median value of \$24.24 indicates the spiky price behavior with large positive outliers. This price pattern is also described by the right skewness with values ranging from 0.6 to 7.5 with a mean of 3.5. The distribution of the spot prices further exhibit large excess kurtosis for all 24 time-series. The statistics on the *forward prices* in table 16.2 are similar to the spot price results. The average forward prices are slightly lower with an average of \$28.67 but exhibit a higher average median value with \$25.75. The standard deviation of \$18.53 is remarkably lower than the spot price volatility showing that forward contracts are more stable over time than spot prices. However, the data sample also contains rather extreme time-series with those of the 6th, 10th, 19th, 20th, and 21th hour which are characterized by a high positive skewness and an excess kurtosis. From these characteristics we see that the Californian electricity prices exhibit a spiky pattern and are highly volatile which was already expected ahead of starting to trade electricity. These findings are supported from the empirical literature

Table 16.1. Sample Statistics of Electricity Spot Prices

Hour	MEAN	STD	MAX	MED	MIN	SK	KU
1	20.66	14.85	107.14	19.03	0.00	1.6	8.9
2	20.10	14.58	134.35	19.36	0.00	1.5	10.0
3	19.35	13.99	99.99	18.51	0.00	1.2	6.8
4	17.82	12.16	72.11	17.00	0.01	0.6	3.3
5	17.36	12.21	75.00	17.50	0.00	0.6	3.4
6	18.99	14.08	148.78	18.52	0.00	2.6	22.7
7	22.88	19.19	172.14	20.65	0.00	2.9	18.6
8	29.11	33.63	434.37	25.00	0.00	6.5	61.1
9	29.21	25.72	331.36	25.50	0.01	5.1	45.7
10	30.30	28.70	360.00	25.76	0.01	6.4	60.4
11	32.56	32.03	492.20	26.79	1.00	7.5	86.4
12	33.49	32.74	414.48	27.00	0.01	6.3	57.5
13	32.66	28.98	250.00	26.04	0.01	4.3	27.1
14	37.26	40.09	250.00	26.70	0.07	3.7	17.7
15	39.19	45.78	250.00	26.90	0.01	3.5	15.2
16	41.28	52.77	274.44	26.00	0.10	3.1	11.8
17	38.48	47.00	250.00	26.05	0.43	3.4	14.9
18	39.85	43.64	308.55	28.37	0.01	3.4	16.2
19	38.46	41.56	252.06	27.85	0.00	3.2	14.7
20	36.25	34.83	256.00	29.00	0.01	3.7	19.9
21	34.48	28.03	250.00	28.82	0.00	4.1	26.0
22	30.82	18.98	152.00	27.84	0.01	2.8	15.6
23	27.96	18.47	242.76	25.89	0.01	4.8	50.2
24	22.33	14.00	117.59	21.65	0.01	1.3	8.4
MEAN	29.62	27.83	237.31	24.24	0.07	3.5	25.9
STD	7.89	12.64	114.34	3.98	0.22	1.9	22.1
MAX	41.28	52.77	492.20	29.00	1.00	7.5	86.4
MIN	17.36	12.16	72.11	17.00	0.00	0.6	3.3

Notes:
Prices are stated in \$/MWh. For each distribution of hourly prices we report the mean, the standard deviation, the maximum, the median, the minimum, the skewness, and the kurtosis. Summary statistics across all time-series are given at the bottom of the table.

as, for example, by Escribano, Pena, and Villaplana (2002) who additionally provide international evidence.[5]

A further examination of the relationship between spot and forward prices is reported in table 16.3. There we analyze the difference between the forward prices of day $t-1$ for electricity delivery on day t with the realized spot prices

[5] The major physical factors for these price patterns are seen in the fuel price, load uncertainty, variations in hydroelectricity production, generation uncertainty, and transmission congestion; see, for example, CalPX (1998).

Table 16.2. Sample Statistics of Electricity Forward Prices

Hour	MEAN	STD	MAX	MED	MIN	SK	KU
1	20.90	9.75	82.00	21.00	0.01	0.7	5.9
2	19.34	9.79	82.00	19.50	0.00	0.8	6.0
3	18.29	9.96	82.00	18.00	0.00	0.8	6.0
4	18.00	9.95	82.00	17.53	0.01	0.8	6.0
5	18.73	10.16	101.35	18.09	0.00	1.1	9.3
6	21.06	13.66	249.00	20.23	0.00	7.4	122.9
7	23.14	13.08	182.51	23.10	0.00	3.3	37.5
8	26.42	13.22	146.59	25.49	1.84	2.5	18.0
9	28.41	15.56	250.00	27.00	0.99	5.9	73.1
10	30.74	30.17	725.00	28.15	4.00	19.3	440.4
11	31.43	15.87	250.00	28.96	4.99	5.4	61.8
12	31.80	16.36	250.00	28.81	4.99	5.0	54.8
13	32.96	18.93	250.00	28.69	4.93	4.3	36.5
14	35.01	21.98	178.65	28.97	4.93	3.0	15.5
15	36.91	26.72	213.13	28.92	4.93	3.1	14.4
16	37.64	29.67	250.00	28.99	3.01	3.3	16.1
17	37.54	27.92	221.27	29.34	4.00	3.1	14.2
18	37.38	24.90	200.20	30.56	4.94	3.0	14.4
19	36.18	33.71	725.00	30.17	4.50	13.9	275.6
20	34.12	32.53	725.00	29.41	4.99	15.5	321.3
21	33.88	30.87	725.00	29.99	6.79	17.8	395.2
22	29.07	10.24	104.02	28.01	6.79	1.7	10.5
23	26.56	10.28	104.60	26.03	4.99	1.4	10.4
24	22.56	9.35	82.10	22.98	3.01	0.7	5.9
MEAN	28.67	18.53	260.89	25.75	3.11	5.2	82.1
STD	6.92	8.70	221.68	4.43	2.39	5.6	131.8
MAX	37.64	33.71	725.00	30.56	6.79	19.3	440.4
MIN	18.00	9.35	82.00	17.53	0.00	0.7	5.9

Notes:
Prices are stated in \$/MWh. For each distribution of hourly prices we report the mean, the standard deviation, the maximum, the median, the minimum, the skewness, and the kurtosis. Summary statistics across all time-series are given at the bottom of the table.

on day t. The left part of the table contains the sample statistics which show that the average spot price lies above the forward price by \$1.10. However, the range of the price differences is rather large with a maximum value of \$714.49 and a low of minus \$426.40 where the medians take values around \$0.94. Further, the right side of table 16.3 reports the results of a regression of the realized day t spot prices against the forward prices of day $t-1$. This enables us to evaluate the question whether the forward prices are biased or unbiased predictors of next day's realized spot prices. The null hypothesis for unbiasedness requires the intercept term and the slope coefficient to be statistically

Table 16.3. Comparison of Electricty Forward and Realized Spot Prices

	Sample Statistics on Difference of Forward Prices - Realized Spot Prices							Regression Results			
								Intercept		Slope	
Hour	MEAN	STD	MAX	MED	MIN	SK	KU	EST	STD	EST	STD
1	-0.01	12.75	57.00	-0.03	-77.80	-1.2	11.3	4.68	1.27	0.78	0.05
2	-0.97	11.91	66.12	0.13	-98.76	-1.3	15.2	3.99	1.13	0.85	0.05
3	-1.30	11.09	49.43	-0.59	-77.05	-1.3	11.6	4.28	1.00	0.84	0.05
4	0.06	10.08	55.54	0.01	-52.81	0.2	6.1	5.45	0.87	0.71	0.04
5	1.16	9.98	71.55	1.03	-39.73	0.7	8.9	4.48	0.86	0.71	0.04
6	1.95	14.29	212.95	1.97	-105.79	4.2	87.4	9.67	0.95	0.47	0.04
7	-0.01	16.35	143.76	1.60	-124.86	-1.2	25.9	5.15	1.40	0.79	0.05
8	-3.00	30.36	73.98	0.21	-381.73	-6.6	66.2	-0.63	2.77	1.14	0.09
9	-0.94	22.94	101.24	0.64	-278.35	-4.6	48.8	7.34	1.90	0.78	0.06
10	0.41	35.89	649.00	1.63	-286.31	7.1	184.0	22.83	1.57	0.25	0.04
11	-1.07	29.35	218.00	1.28	-426.40	-6.4	90.5	6.55	2.57	0.83	0.07
12	-1.83	30.02	218.11	1.07	-339.61	-5.1	57.8	7.49	2.60	0.82	0.07
13	0.15	26.62	215.76	1.40	-210.10	-1.9	25.4	10.33	2.07	0.68	0.05
14	-2.27	34.60	137.92	1.48	-217.23	-2.6	15.9	4.85	2.58	0.93	0.06
15	-2.33	40.85	164.40	1.81	-219.80	-2.4	14.7	9.66	2.75	0.80	0.06
16	-3.71	46.43	192.86	2.50	-220.92	-2.0	11.4	8.93	2.97	0.86	0.06
17	-0.97	40.69	177.85	3.00	-218.41	-2.2	13.7	6.35	2.69	0.86	0.06
18	-2.67	37.03	158.87	1.91	-263.25	-2.2	14.8	5.16	2.67	0.93	0.06
19	-2.54	46.79	714.49	2.34	-220.00	4.3	93.9	28.07	2.37	0.30	0.05
20	-2.28	40.82	671.09	0.60	-224.31	5.7	124.2	26.78	1.94	0.29	0.04
21	-0.73	37.34	705.30	0.20	-211.62	9.7	207.1	28.60	1.62	0.18	0.04
22	-1.88	17.47	70.43	-0.81	-114.65	-1.9	13.0	9.24	2.09	0.75	0.07
23	-1.48	17.12	62.00	-1.10	-200.77	-4.3	51.5	8.97	1.88	0.72	0.07
24	-0.11	11.42	47.17	0.17	-72.59	-0.8	8.0	3.89	1.23	0.84	0.05
MEAN	-1.10	26.34	218.12	0.94	-195.12	-0.7	50.3	9.67	1.91	0.71	0.05
STD	1.38	12.65	221.70	1.08	104.09	4.1	56.0	8.16	0.70	0.24	0.01
MAX	1.95	46.79	714.49	3.00	-39.73	9.7	207.1	28.60	2.97	1.14	0.09
MIN	-3.71	9.98	47.17	-1.10	-426.40	-6.6	6.1	-0.63	0.86	0.18	0.04

Notes:
The price differences are stated in \$/MWh and their distribution is characterized by values for the mean, the standard deviation, the maximum, the median, the minimum, the skewness, and the kurtosis. Further, we report results on a ordinary least squares regression of forward prices on spot prices. At the bottom of the table we show summary statistics across all time-series.

not different from zero and one, respectively. Looking at the results indicates that the null hypothesis can be rejected at conventional significance levels for almost all 24 time-series. The intercept term lies above zero at an average of 9.67 with a corresponding standard error of 1.91. The slope coefficients show values that are almost all significantly below one with an average of 0.71 at a standard deviation of 0.05. Thus, even for a time period of only one-day ahead the forward prices are not an unbiased predictor of realized spot prices which demonstrates the need for sound theoretical valuation models. From a theoretical pricing point of view the equilibrium model of Bessembinder and Lemmon (2002) implies that the forward electricity price is a downward biased predictor of the future spot price, if expected power demand is low and demand risk is moderate. However, they find the equilibrium forward premium increases when either expected demand or demand variance is high, because of the positive skewness in the electricity spot price distribution.

In order to capture the spot price behavior and to price the day-ahead forward contracts we select two theoretical pricing models for *calibration* to the given electricity data. *First*, we choose to implement the one-factor model as parametrized in Schwartz (1997, Model 1) which originally is intended to capture the price processes on oil spot and futures markets. We, however, use this equilibrium pricing model to describe the electricity spot price behavior by the continuous-time process

$$dS = \kappa\left(\mu - \ln S\right) S dt + \sigma S dz^{\mathbb{P}} \tag{16.17}$$

yielding a corresponding arbitrage-free log-forward price of

$$\begin{aligned} \ln F\left(t,T\right) &= e^{-\kappa(T-t)} \ln S_t + \alpha^* \left(1 - e^{-\kappa(T-t)}\right) + \frac{\sigma^2}{4\kappa}\left(1 - e^{-2\kappa(T-t)}\right), \\ \text{with } \alpha^* &= \mu - \frac{\sigma^2}{2\kappa} - \lambda \end{aligned} \tag{16.18}$$

under the martingale measure $\mathbb{Q}$ including the time invariant market price of risk parameter λ. Based on equations (16.17) and (16.18) as transition and measurement equation, respectively, we estimate the model using the standard Kalman filter setup. The results on the parameter estimates including the two measurement errors $\sigma_{\varepsilon,S}$ and $\sigma_{\varepsilon,F}$ are shown in tables 16.4 and 16.5. The coefficient of mean-reversion is estimated by 0.6 on average across all 24 time-series. The values translate to mean half-lives of the electricity spot price with a minimum of 0.6 years for hour 16 and a maximum period of 3.8 years for the 22nd hour where the average lies at 1.1 years. The rather not statistically significant estimates for the long-run mean μ reach values around

Table 16.4. Parameter Estimates for Schwartz (1997, Model 1)

	κ		μ		σ	
Hour	EST	STD	EST	STD	EST	STD
1	0.287***	0.005	2.516	2.351	2.405***	0.381
2	0.518***	0.042	2.569**	1.187	3.019***	0.202
3	0.724***	0.073	2.498	3.854	3.060***	0.415
4	0.370***	0.014	2.536**	1.131	2.883***	0.352
5	0.770***	0.077	2.561***	0.388	3.060***	0.532
6	0.611***	0.237	2.444	3.373	2.668***	0.557
7	0.645	0.690	2.871	17.650	2.746***	0.536
8	0.294	0.390	2.501	60.805	2.161***	0.830
9	1.022	1.291	3.451	27.236	4.126**	2.062
10	0.341***	0.062	2.743	61.953	2.383***	0.881
11	0.467***	0.102	2.829	24.289	2.810**	1.185
12	0.445***	0.138	2.873	10.067	3.042***	0.656
13	0.483***	0.088	3.299	6.596	3.136***	0.440
14	0.901***	0.080	3.250	9.597	4.026***	0.412
15	1.101***	0.311	3.505	12.857	4.526***	0.364
16	1.170**	0.569	3.371	12.982	4.629***	0.407
17	1.076***	0.075	3.238	11.428	4.391***	0.370
18	0.972***	0.199	3.120	6.447	4.337***	0.392
19	0.752***	0.256	3.134	2.480	3.815***	0.485
20	0.458***	0.072	2.600	5.658	2.992***	0.451
21	0.249***	0.011	2.651	7.278	2.459***	0.404
22	0.182***	0.030	2.292	28.804	2.319***	0.208
23	0.289***	0.073	2.400	26.339	2.615***	0.666
24	0.351***	0.035	2.624	2.495	2.561***	0.460
MEAN	0.603	0.205	2.828	14.469	3.174	0.569
STD	0.304	0.291	0.372	16.879	0.773	0.385
MAX	1.170	1.291	3.505	61.953	4.629	2.062
MIN	0.182	0.005	2.292	0.388	2.161	0.202

Note:
Statistically significant parameter estimates at the 1%-, 5%- and 10%-levels are denoted by ***, **, and *, respectively.

2.8. Related to the coefficient of mean-reversion is the σ-parameter which exhibits how volatile the spot prices fluctuate around their long-run mean. At statistically significant levels the volatility parameter takes values of 216 to 463 percent p.a. which underlines the characteristic of extreme volatility patterns in electricity prices. All volatility point estimates are statistically significant at the one percent level, except for hour 11. Further, the market price of risk parameter λ is estimated at a mean value of approximately -1.0 across all 24 time-series. These negative inferred parameter values for the market price of risk can be seen as a compensation of the market participants

Table 16.5. Parameter Estimates for Schwartz (1997, Model 1) (continued)

	λ		$\sigma_{\varepsilon,S}$		$\sigma_{\varepsilon,F}$		neg.
Hour	EST	STD	EST	STD	EST	STD	LogL
1	-0.921*	0.513	0.650***	0.033	0.181***	0.018	653.2
2	-1.087***	0.268	0.641***	0.033	0.169***	0.001	669.4
3	-0.955	0.774	0.574***	0.029	0.232***	0.029	727.3
4	-1.131***	0.064	0.634***	0.034	0.276***	0.032	850.9
5	-0.943***	0.333	0.675***	0.036	0.258***	0.031	865.2
6	-0.829	0.740	0.613***	0.028	0.303***	0.034	862.1
7	-0.739	5.129	0.619***	0.033	0.367***	0.030	965.7
8	-0.753	8.213	0.634***	0.033	0.331***	0.031	922.9
9	-1.237	6.387	0.638***	0.037	0.257***	0.081	933.7
10	-0.820	9.350	0.544***	0.033	0.237***	0.038	690.3
11	-0.953	4.871	0.525***	0.034	0.193***	0.058	612.8
12	-1.110	1.264	0.510***	0.040	0.173***	0.035	568.2
13	-1.075	0.752	0.466***	0.028	0.200***	0.027	578.4
14	-1.310	2.103	0.572***	0.036	0.187***	0.026	751.8
15	-1.435	2.816	0.557***	0.031	0.181***	0.027	763.7
16	-1.494	3.568	0.608***	0.031	0.194***	0.028	845.3
17	-1.444	2.674	0.591***	0.031	0.173***	0.029	777.0
18	-1.498	1.505	0.635***	0.032	0.142***	0.026	753.3
19	-1.318***	0.215	0.687***	0.034	0.182***	0.035	826.4
20	-1.112	0.783	0.600***	0.030	0.192***	0.031	700.0
21	-0.969*	0.519	0.532***	0.024	0.177***	0.029	547.6
22	-0.983	2.302	0.539***	0.031	0.111***	0.013	361.7
23	-1.050	3.240	0.618***	0.038	0.089**	0.04	422.7
24	-0.940	0.598	0.622***	0.027	0.107***	0.026	458.9
MEAN	-1.088	2.458	0.595	0.032	0.205	0.032	712.9
STD	0.230	2.605	0.055	0.004	0.068	0.015	163.7
MAX	-0.739	9.350	0.687	0.040	0.367	0.081	361.7
MIN	-1.498	0.064	0.466	0.024	0.089	0.001	965.7

Note:
Statistically significant parameter estimates at the 1%-, 5%- and 10%-levels are denoted by ***, **, and *, respectively.

for the high positive skewness of the spot prices when forward contracts are to be traded on such an underlying. Finally, the measurement errors give evidence that the spot electricity prices on average are only captured by a three times higher standard deviation of 0.60 than the 0.20 estimated for the forward prices. Overall, the information we get from calibrating the forward pricing model of Schwartz (1997, Model 1) to the electricity data is that the spot prices follow a mean-reverting process which is highly volatile, and the cross-section combining the spot and the forward prices reveals negative values for the market price for unhedgeable spot price movements.

Table 16.6. Inference Results for the Stochastic Volatility Model

	μ		λ		κ		σ	
Hour	EST	STD	EST	STD	EST	STD	EST	STD
1	1.865	1.375	-10.739***	2.183	1.156***	0.253	1.815***	0.635
2	0.153	0.813	-4.753***	1.289	7.359***	1.542	9.476***	2.053
3	-0.311	1.183	-3.298**	1.247	1.607***	0.165	3.773***	0.761
4	-0.092	0.446	-3.486	2.205	2.148	2.351	2.614	3.175
5	3.466	2.367	-8.236***	1.243	0.434	1.994	0.319	0.221
6	-0.093	0.605	-11.086***	2.862	0.982	1.819	0.686	1.276
7	-0.534	1.621	-13.006***	2.691	5.664***	1.698	7.107***	1.454
8	0.010	0.037	-7.490***	2.464	2.279*	1.211	3.301***	0.761
9	-0.982	1.512	-2.131***	0.074	0.620***	0.101	3.423***	0.679
10	0.535	1.365	-1.788	2.411	1.760	1.862	0.678	0.513
11	-0.248	2.837	-7.294***	1.495	2.069**	1.039	2.772***	0.645
12	2.067	1.226	-6.055***	1.181	1.092**	0.477	3.871***	0.888
13	0.108	0.371	-5.549***	1.047	1.424***	0.488	4.222***	0.706
14	-0.072	0.309	-3.445***	0.602	1.535***	0.461	6.240***	0.985
15	0.051	0.229	-1.348***	0.403	0.944	2.382	0.281	0.407
16	4.103**	2.129	-3.175***	0.527	0.894*	0.469	4.488***	0.677
17	4.562*	2.546	-4.153***	0.724	0.399	0.587	2.487***	0.639
18	7.266***	2.185	-4.991***	0.958	4.297***	1.006	6.757***	1.034
19	3.440*	2.056	-6.187***	1.178	3.763***	0.919	5.896***	0.949
20	0.347	3.188	-4.282***	1.101	2.638*	1.462	3.112***	1.110
21	0.720	1.519	-2.517***	0.697	0.803	0.496	1.185***	0.264
22	0.006	0.068	-1.824	2.259	0.229	0.297	1.244***	0.403
23	0.115	0.473	-1.107	1.797	0.937	2.019	0.285	0.298
24	0.019	0.070	-1.860	2.167	0.776	1.490	0.970	0.615
MEAN	1.104	1.272	-4.992	1.450	1.909	1.108	3.208	0.881
STD	2.034	0.947	3.259	0.789	1.752	0.731	2.478	0.637
MAX	7.266	3.188	-1.107	2.862	7.359	2.382	9.476	3.175
MIN	-0.982	0.037	-13.006	0.074	0.229	0.101	0.281	0.221

Note:
Statistically significant parameter estimates at the 1%-, 5%- and 10%-levels are denoted by ***, **, and *, respectively.

Second, we estimate our theoretical stochastic volatility forward pricing model derived in section 15.2 using the setup for the empirical inferences as described in section 16.1. The results on the model parameter estimates are shown in tables 16.6 and 16.7. In the estimation procedure we set the market price of risk parameter λ_v for the variance rate equal to zero and use the squared point estimates of the volatility parameter σ from the empirical inferences on Schwartz (1997, Model 1) as input for the long-run mean θ of the variance rate. Thus, with the parameter space of interest given by $\psi = \{\mu, \lambda, \kappa, \sigma, \rho, \sigma_{\varepsilon,S}, \sigma_{\varepsilon,F}\}$ we have one state variable and one parameter

Table 16.7. Inference Results for the Stochastic Volatility Model (continued)

	ρ		$\sigma_{\varepsilon,S}$		$\sigma_{\varepsilon,F}$		neg.
Hour	EST	STD	EST	STD	EST	STD	LogL
1	-0.646***	0.132	0.640***	0.019	0.183***	0.008	640.3
2	-0.432***	0.106	0.619***	0.020	0.173***	0.010	640.4
3	-0.697***	0.082	0.554***	0.018	0.245***	0.012	701.3
4	-0.757	0.580	0.628***	0.020	0.285***	0.015	846.1
5	-0.467	0.795	0.668***	0.021	0.261***	0.011	859.0
6	-0.952***	0.034	0.608***	0.023	0.310***	0.014	852.4
7	-0.333**	0.157	0.561***	0.018	0.360***	0.014	901.2
8	-0.596**	0.258	0.625***	0.020	0.336***	0.012	908.9
9	-0.657***	0.118	0.639***	0.018	0.275***	0.011	903.6
10	-0.606	0.963	0.553***	0.019	0.238***	0.010	697.7
11	-0.032	0.149	0.507***	0.015	0.191***	0.009	580.5
12	0.868***	0.065	0.506***	0.015	0.178***	0.009	532.0
13	0.654***	0.091	0.452***	0.014	0.201***	0.010	539.3
14	0.654***	0.114	0.554***	0.017	0.164***	0.011	674.0
15	0.111	0.235	0.558***	0.017	0.179***	0.011	761.3
16	0.802***	0.070	0.594***	0.018	0.205***	0.011	802.6
17	0.146	0.447	0.580***	0.018	0.188***	0.011	746.2
18	0.854***	0.059	0.616***	0.018	0.148***	0.010	703.3
19	0.721***	0.099	0.664***	0.020	0.179***	0.011	785.0
20	0.023	0.131	0.582***	0.019	0.195***	0.009	677.3
21	-0.033	0.056	0.531***	0.016	0.176***	0.008	538.9
22	-0.855**	0.427	0.535***	0.016	0.122***	0.007	345.7
23	-0.429	0.405	0.619***	0.018	0.090***	0.006	419.0
24	-0.999**	0.420	0.627***	0.019	0.118***	0.008	470.1
MEAN	-0.152	0.250	0.584	0.018	0.208	0.010	688.6
STD	0.624	0.247	0.054	0.002	0.068	0.002	158.2
MAX	0.868	0.963	0.668	0.023	0.360	0.015	908.9
MIN	-0.999	0.034	0.452	0.014	0.090	0.006	345.7

Note:
Statistically significant parameter estimates at the 1%-, 5%- and 10%-levels are denoted by ***, **, and *, respectively.

more to calibrate our model to the spot and forward electricity prices as compared to the single-factor model of Schwartz (1997, Model 1). The higher flexibility of the stochastic volatility model is underlined by a lower measurement error $\sigma_{\varepsilon,S}$ in the spot prices averaging at 0.58 instead of 0.60 as with the single-factor model. This gives evidence that the time-series behavior of the electricity data is captured closer when we include the variance rate as second factor. The cross-sectional fit of the term structure calibrated by the second measurement equation of (16.15), however, is not improved by the two-factor model; the measurement error $\sigma_{\varepsilon,F}$ stays approximately unchanged at a value of 0.21. Next, we look at the other parameter estimates. The expected rate of return μ of the spot prices is estimated at levels ranging from -0.98 up to 7.2. As expected for a level parameter, the inferences on the mean rate are not statistically different from zero. The values for the market price of spot price risk λ are deduced from the data at a negative spectrum starting from -1.1 and ending at a value of -13.0. With the stochastic volatility pricing model the market prices of risk are mostly estimated at significant negative levels. The inferred negative values are in line with the results from the estimation of Schwartz (1997, Model 1). The highly negative parameter values for the market price of risk further demonstrate that it is crucial to calibrate an equilibrium forward pricing model (see, for example, Bessembinder and Lemmon (2002)) in that the non-storable commodity of electricity creates unhedgeable risk. This risk need to be compensated for on the short sellers' side of the forward contracts. Further evidence that premia in electricity forward prices are related to measures of risks faced by market participants is provided by the papers of Longstaff and Wang (2002) and Longstaff and Wang (2003), which give deeper insight into the determinants of energy derivative prices. The remaining parameters κ, σ, and ρ describe the time-series behavior of the second state variable, i.e. the variance rate. The variance rate is estimated with high tendencies to revert to its long-run equilibrium value; the inferences on the coefficients of mean-reversion κ translate to mean half-lives with a minimum of 1.1 months and a maximum of 36.3 months. The volatility of the variance rate is inferred at generally highly significant values with an average of 320 percent across all 24 time series. Finally, the correlation coefficient ρ, which shows whether the two state variables generally move in the same or in the opposite direction, is estimated at a negative average value of -0.15. Only 37.5 percent or 9 out of the 24 time-series exhibit correlation coefficients above zero.

17

Summary and Conclusions

In part IV of this book we focus on pricing short-term electricity forward contracts. Based on the peculiarities of electricity as underlying commodity of forward contracts we develop a theoretical valuation model. In analyzing electricity spot and forward market data we further present results on empirical inferences for the Californian power market.

In the introductory chapter we investigate the more mature markets of commodity futures to clear the facts on the pricing of electricity forwards. Especially the fact of non-storability of electricity is found to be important in that it causes extreme spikes and high volatility in the electricity spot prices and asks for pricing in an incomplete market situation. In achieving to capture such price behavior of electricity we choose to build a stochastic factor model to price electricity forwards as described in chapter 15. We use the non-tradeable spot price of electricity and the variance rate of the spot prices as underlying state variables including their market prices of risk. Thereupon, we present a closed-form solution on valuing electricity forward contracts using risk-neutral pricing techniques.

In chapter 16 we then show how to implement the theoretical pricing model on empirical data. There, we first clear distributional issues of the stochastic state variables and discretize the continuous-time pricing model into a convenient state space model. Grounded on the state space representation we show how to run an extended Kalman filter algorithm in order to calibrate our pricing model to empirical spot and forward data on electricity by means of maximum likelihood inferences. For the historical sample we use hourly spot and day-ahead forward electricity data from the largely deregulated Californian power market available for the period of April 1998 to December 1999. In describing the data sample we find the spot and forward prices to average around $29 per MWh and exhibit high standard deviations. The distribution of the spot prices is further characterized by a high positive skewness and an excess kurtosis. The findings for the forward prices are similar, but some time series exhibit even higher values for the skewness and kurtosis than they do in the spot prices. Both the spot and the forward time series show extreme spikes in prices and a highly volatile price behavior. Further regression results provide evidence that the day-ahead forward prices are not unbiased predictors for realized spot prices. This creates the need for theoretical valuation models to price electricity forward contracts. We choose to calibrate a single-factor model and our stochastic volatility pricing model to the data. In comparing the two models, we find the higher flexibility of the stochastic volatility model resulting especially in a better fit of the time-series behavior of the spot prices. Furthermore, the inferences on the model parameters primarily provide evidence for the need of choosing an equilibrium valuation model to price electricity forwards. This result stems from largely negative estimates for the market prices of unhedgeable spot electricity price risk in both models. Finally, it would have been interesting to calibrate a two-factor model that captures the mean-reversion property in both state variables.

List of Symbols and Notation

The symbols and notation used throughout this study are introduced at their first appearance. The choices of symbols and notation are adopted from common uses in the literature. In the following we enumerate and explain the most frequently used symbols and notation separately for each part of the book.

Part I

A, B	factor loadings
$B(0,t)$	money market account
$\mathbb{C}ov[\ldots\vert\ldots]$	conditional covariance operator
dt	infinitesimal time increment
dW_t	infinitesimal Brownian increment
$\mathbb{E}[\ldots\vert\ldots]$	conditional expectation operator
ε_t	vector of measurement errors
$\boldsymbol{\eta}_t$	vector of transition equation noise terms
$\mathbf{F}_t$	variance-covariance matrix of the prediction error
$F(\ldots)$	distribution function
$\mathcal{F}_t$	information set up to time t
$\mathbf{g}_t$	general non-linear measurement function
$\mathbf{h}_t$	general non-linear transition function
$\mathbf{K}_t$	Kalman gain matrix
$L(\ldots)$	log-likelihood function
λ	market price of risk
$\mathbf{M}_t$	martingale process
$MMSE$	minimum mean square error
$MMSLE$	minimum mean square linear error
$\mathbb{MSE}(\ldots)$	mean square error
$N(\ldots)$	normal distribution function

$p(\ldots)$	density function
$P(t,T)$	price of a discount bond
$\mathbb{P}$	empirical probability measure
$\boldsymbol{\psi}$	vector of parameters
$\mathbf{S}_t$	stock price
$\mathbf{S}_t^*$	discounted stock price
Σ_t	variance-covariance matrix of the state variables
t	variable for calendar time
$\mathbb{V}ar[\ldots\vert\ldots]$	conditional variance operator
$\mathbf{X}_t$	general stochastic process
$\boldsymbol{\xi}_t$	vector of state variables
$\mathbf{y}_t$	vector of measurable observations

Part II

A, B, C	factor loadings
$AR_{t,i}$	abnormal return of security i on date t
BM_t	benchmark return
CAR	cumulated abnormal return
dt	infinitesimal time increment
dW_t	infinitesimal Brownian increment
$\mathbb{E}[\ldots\vert\ldots]$	conditional expectation operator
$HP_{t,i}$	holding period return of security i on date t
κ	speed of mean-reversion
KU	kurtosis
$L(\ldots)$	log-likelihood function
MAD	mean absolute deviation
μ	expected mean return of the net asset value
NAV_t	net asset value of closed-end funds
P_t	market price of closed-end funds
π_t	instantaneous dynamic premium
$PREM_t$	empirical premium
$\boldsymbol{\psi}$	vector of parameters
r	riskless interest rate
θ	long-run mean value
$\mathbb{V}ar[\ldots\vert\ldots]$	conditional variance operator
X_t	logarithm of the net asset value
$\boldsymbol{\xi}_t$	vector of state variables
Y_t	logarithm of the market price

Part III

A, B, C	factor loadings
$B(0,t)$	money market account
$\mathbb{C}ov[\ldots\vert\ldots]$	conditional covariance operator
dt	infinitesimal time increment
dW_t	infinitesimal Brownian increment
$\mathbb{E}[\ldots\vert\ldots]$	conditional expectation operator
$f(t,T)$	forward interest rate
$\mathcal{F}_t$	information set up to time t
$k(t,T)$	swap rate
KU	kurtosis
$L(t,T)$	LIBOR rate
λ	market price of risk
$P(t,T)$	price of a discount bond
$\mathbb{P}$	empirical probability measure
r	instantaneous interest rate
ρ	coefficient of correlation
$\mathbb{Q}$	risk neutral probability measure
$\mathbb{Q}^{\mathrm{T}}$	forward martingale measure
SK	skewness
STD	standard deviation
$\mathbb{V}ar[\ldots\vert\ldots]$	conditional variance operator
$\mathbf{x}$	vector of theoretical state variables
$\boldsymbol{\xi}_t$	vector of empirical state variables
$\mathbf{y}_t$	vector of measurable observations
$y(t,T)$	spot interest rate
Y_t^T	forward price of a discount bond

Part IV

$\mathbb{C}ov[\ldots\vert\ldots]$	conditional covariance operator
dt	infinitesimal time increment
dW_t	infinitesimal Brownian increment
$\mathbb{E}[\ldots\vert\ldots]$	conditional expectation operator
$F(t,T)$	forward or futures price
$\mathcal{F}_t$	information set up to time t
κ	speed of mean-reversion
KU	kurtosis
$L(\ldots)$	log-likelihood function
λ	market price of risk

μ	expected mean return
$P(t,T)$	price of a discount bond
$\mathbb{P}$	empirical probability measure
$\mathbb{Q}$	risk neutral probability measure
ρ	coefficient of correlation
S_t	spot price
σ	instantaneous volatility
SK	skewness
STD	standard deviation
θ	long-run mean volatility
v_t	stochastic variance rate
$\mathbb{V}ar[\ldots\|\ldots]$	conditional variance operator
X_t	logarithm of spot price
$\boldsymbol{\xi}_t$	vector of empirical state variables
$\mathbf{y}_t$	vector of measurable observations

List of Tables

List of Figures

References

ABRAMOWITZ, M., AND I. A. STEGUN (1965): *Handbook of Matehmatical Functions with Formulas, Graphs and Mathematical Tables.* New York: Dover Publications, 9 edn.

AIT-SAHALIA, Y. (1996): "Testing Continuous-Time Models of the Spot Interest Rate," *Review of Financial Studies*, 9(2), 385–426.

AITCHISON, J., AND J. A. C. BROWN (1957): *The Lognormal Distribution with Special References to Its Uses in Economics.* Cambridge: Cambridge University Press.

AMIN, K., V. NG, AND S. C. PIRRONG (1995): "Valuing Energy Derivatives," in *Managing Energy Price Risk*, ed. by R. Jamson, pp. 57–70. London: Risk Publications.

ANDERSEN, T. G., AND J. LUND (1997): "Estimating Continuous-Time Stochastic Volatility Models of the Short-Term Interest Rate," *Journal of Econometrics*, 77(2), 343–377.

ANDERSON, B. D. O., AND J. B. MOORE (1979): *Optimal Filtering.* Englewood Cliffs: Prentice Hall.

ANDERSON, S. C. (1986): "Closed-End Funds versus Market Efficiency," *Journal of Portfolio Management*, 13(1), 63–65.

ANDERSON, S. C., AND J. A. BORN (1992): *Closed-End Investment Companies: Issues and Answers.* Boston: Kluwer.

——— (2002): *Closed-End Fund Pricing: Theory and Evidence.* Boston: Kluwer.

AOKI, M. (1990): *State Space Modeling of Time Series.* Berlin: Springer, 2 edn.

ARORA, N., N. JU, AND H. OU-YANG (2002): "Closed-End Funds: A Dynamic Model of Premiums and Discounts," Working paper, University of North Carolina.

ARTZNER, P. (1997): "On the Numeraire Portfolio," in *Mathematics of Derivative Securities*, ed. by M. A. Dempster, and S. R. Pliska, pp. 53–58. Cambridge: Cambridge University Press.

AVELLANEDA, M., AND P. LAURENCE (2000): *Quantitative Modeling of Derivative Securities - From Theory to Practice.* Boca Raton: Chapman & Hall.

BABBS, S. H., AND K. B. NOWMAN (1999): "Kalman Filtering of Generalized Vasicek Term Structure Models," *Journal of Financial and Quantitative Analysis*, 34(1), 115–130.

BACHELIER, L. (1900): "Thorie de la Spculation," in *The Random Character of Stock Market Prices (English translation, 1964)*, ed. by P. H. Cootner, pp. 17–78. Cambridge: MIT Press.

BACKUS, D., S. FORESI, AND S. ZIN (1998): "Arbitrage Opportunities in Arbitrage-Free Models on Bond Pricing," *Journal of Business and Economic Statistics*, 16(1), 13–26.

BALL, C. A., AND W. N. TOROUS (1996): "Unit Roots and the Estimation of Interest Rate Dynamics," *Journal of Empirical Finance*, 3(2), 215–238.

BARBER, P. (1997): "Rationalising the Transmission Business," in *The US Power Market*, ed. by R. Jameson. London: Risk Publications.

BATES, D. S. (1996): "Testing Option Pricing Models," in *Handbook of Statistics*, ed. by G. Maddala, and C. R. Rao, vol. 14, pp. 567–611. Amsterdam: Elsevier.

BEAGLEHOLE, D. R., AND M. S. TENNEY (1991): "General Solutions of Some Interest Rate-Contingent Claim Pricing Equations," *Journal of Fixed Income*, 1(9), 69–83.

BELL, J. C., AND J. R. LILYESTROM (1997): "The March of the Regulator: Regulatory Dynamics at the State and Federal Level," in *The US Power Market*, ed. by R. Jameson, pp. 17–32. London: Risk Publications.

BERGSTROM, A. (1984): "Continuous Time Stochastic Models and Issues of Aggregation over Time," in *Handbook of Econometrics*, ed. by Z. Griliches, and M. Intriligator, vol. 2, pp. 1145–1212. Amsterdam: Elsevier.

BESSEMBINDER, H., AND M. L. LEMMON (2002): "Equilibrium Pricing and Optimal Hedging in Electricity Forward Markets," *Journal of Finance*, 57(3), 1347–1382.

BESTER, C. A. (2003): "Random Field Models and Affine Models for Interest Rates: An Empirical Comparison," Working paper, Duke University.

BHANOT, K. (2000): "Behavior of Power Prices: Implications for the valuation and Hedging of Financial Contracts," *Journal of Risk*, 2(3), 43–62.

BIELECKI, T. R., AND M. RUTKOWSKI (2001): *Credit Risk: Modelling, Valuation and Hedging*. New York: Springer.

BIERMAN, H., AND B. SWAMINATHAN (2000): "Managing a Closed-End Investment Fund," *Journal of Portfolio Mangement*, 26(4), 49–52.

BIERWAG, G., G. C. KAUFMAN, AND A. TOEVS (1983): "Innovations in Bond Portfolio Management: Duration Analysis and Immunization," Greenwich: JAI Press.

BJERRING, J. H., J. LAKONISHOK, AND T. VERMAELEN (1983): "Stock Prices and Financial Analysts' Recommendations," *Journal of Finance*, 38(1), 187–204.

BJÖRK, T. (1997): "Interest Rate Theory," in *Financial Mathematics*, ed. by W. J. Runggaldier, pp. 53–122. Berlin: Springer.

BLACK, F., AND M. SCHOLES (1973): "The Pricing of Options and Corporate Liabilities," *Journal of Political Economy*, 81(3), 637–654.

BLISS, R. R. (1997): "Testing Term Structure Estimation Methods," *Advances in Futures and Options Research*, 9, 197–231.

——— (1999): "Bliss Term Structures," Technical report, Federal Reserve Bank of Chicago.

BLISS, R. R., AND E. I. RONN (1998): "Callable U.S. Treasury Bonds: Optimal Calls, Anomalies, and Implied Volatilities," *Journal of Business*, 71(2), 211–251.

BOLLERSLEV, T., AND J. M. WOOLDRIDGE (1992): "Quasi-Maximum Likelihood Estimation and Inference in Dynamic Models with Time-Varying Covariances," *Econometric Reviews*, 11(2), 143–172.

BONSER-NEAL, C., G. BRAUER, R. NEAL, AND S. WHEATLEY (1990): "International Investment Restrictions and Closed-End Country Fund Prices," *Journal of Finance*, 45(2), 523–547.

BORENSTEIN, S. (2002): "The Trouble with Electricity Markets: Understanding California's Restructuring Disaster," *Journal of Economic Perspectives*, 16(1), 191–211.

BRACE, A., D. GATAREK, AND M. MUSIELA (1997): "The Market Model of Interest Rate Dynamics," *Mathematical Finance*, 7(2), 127–154.

BRAUER, G. A. (1984): "'Open-Ending' Closed-End Funds," *Journal of Financial Economics*, 13, 491–507.

——— (1988): "Closed-End Fund Shares' Abnormal Returns and the Information Content of Discounts and Premiums," *Journal of Finance*, 43(1), 113–127.

BRENNAN, M. J. (1991): "The Price of Convenience and the Valuation of Commodity Contingent Claims," in *Stochastic Models and Option Values*, ed. by D. Lund, and B. Oksendal, pp. 33–71. Elsevier: Amsterdam.

BRENNAN, M. J., AND E. S. SCHWARTZ (1978): "Finite Difference Methods and Jump Processes Arising in the Pricing of Contingent Claims: A Synthesis," *Journal of Financial and Quantitative Analysis*, 13, 462–474.

——— (1979): "A Continuous Time Approach to the Pricing of Bonds," *Journal of Banking and Finance*, 3(2), 133–155.

——— (1980): "Conditional Predictions of Bond Prices and Returns," *The Journal of Finance*, 35(2), 405–419.

——— (1982): "An Equilibrium Model of Bond Pricing and a Test of Market Efficiency," *Journal of Financial and Quantitative Analysis*, 17(3), 301–329.

——— (1985): "Evaluating Natural Resource Investments," *Journal of Business*, 58(2), 135–157.

BRICKLEY, J. A., AND J. S. SCHALLHEIM (1985): "Lifting the Lid on Closed-End Investment Companies: A Case of Abnormal Returns," *Journal of Financial and Quantitative Analysis*, 20(1), 107–117.

BRIGO, D., AND F. MERCURIO (2001): *Interest Rate Models - Theory and Praxis*. New York: Springer.

BRIYS, E., M. BELLALAH, H. M. MAI, AND F. DE VARENNE (1998): *Options, Futures and Exotic Derivatives - Theory, Application and Practice*. Chichester: John Whiley & Sons.

BRIYS, E., M. CROUHY, AND R. SCHÖBEL (1991): "The Pricing of Default-free Interest Rate Cap, Floor, and Collar Agreements," *Journal of Finance*, 46(5), 1879–1892.

BROCK, W. A., AND A. G. MALLIARIS (1989): *Differential Equations, Stability, and Chaos in Dynamic Eonomics.* Amsterdam: North-Holland.

BROCKWELL, P. A., AND R. A. DAVIES (1987): *Time Series: Theory and Models.* New York: Springer.

BROWN, R. H., AND S. M. SCHAEFER (1994): "The Term Structure of Real Interest Rates and the Cox, Ingersoll, and Ross Model," *Journal of Financial Economics*, 35(1), 3–42.

BROWN, S. J., AND P. H. DYBVIG (1986): "The Empirical Implications of the Cox, Ingersoll, Ross Theory of the Term Structure of Interest Rates," *Journal of Finance*, 41(3), 616–628.

BROZE, L., O. SCAILLET, AND J.-M. ZAKOIAN (1995): "Testing for Continuous-Time Models of the Short-Term Interest Rate," *Journal of Empirical Finance*, 2, 199–223.

BÜHLER, W., O. KORN, AND R. SCHÖBEL (2000): "Pricing and Hedging of Oil Futures - A Unifying Approach," Discussion paper, University of Tübingen.

BURRIDGE, P., AND K. F. WALLIS (1988): "Prediction Theory for Autoregressive Moving Average Processes," *Econometric Reviews*, 7(1), 65–95.

CALPX (1998): *Great Expectations: What Happens When the Markets Open?* Pasadena: California Power Exchange.

——— (1999): *CalPX Primer.* Pasadena: California Power Exchange.

——— (2000a): *1998-1999 Market Year Report to Californians.* Pasadena: California Power Exchange.

——— (2000b): *Providing the Efficient, Competitive Marketplace for Electricity Trading in California.* Pasadena: California Power Exchange.

CAMPBELL, J. Y., A. W. LO, AND A. C. MACKINLAY (1997): *The Econometrics of Financial Markets.* Princeton: Princeton University Press.

CAMPBELL, J. Y., AND R. J. SHILLER (1987): "Cointegration and Tests of Present Value Models," *Journal of Political Economy*, 95, 1062–1088.

CASASSUS, J., AND P. COLLIN-DUFRESNE (2002): "'Maximal' Affine Model of Convenience Yields Implied from Interest Rates and Commodity Futures," Working paper, Carnegie Mellon University.

CHACKO, G., AND S. DAS (2002): "Pricing Interest Rate Derivatives: A Deneral Approach," *Review of Financial Studies*, 15, 195–241.

CHAN, K. C., A. KAROLYI, F. A. LONGSTAFF, AND A. B. SANDERS (1992): "An Empirical Comparison of Alternative Models of the Short-Term Interest Rate," *Journal of Finance*, 47(3), 1209–1225.

CHAPMAN, D. A., J. B. LONG, AND N. D. PEARSON (1999): "Using Proxies for the Short Rate: When are Months like an Instant?," *Review of Financial Studies*, 12(4), 763–806.

CHAPMAN, D. A., AND N. D. PEARSON (2001): "Recent Advances in Estimating Term-Structure Models," *Financial Analysts Journal*, 57(7&8), 77–95.

CHEN, L. (1996a): *Interest Rate Dynamics, Derivatives Pricing, and Risk Management.* Berlin: Springer.

CHEN, N.-F., R. KAN, AND M. H. MILLER (1993): "Are the Discounts on Closed-End Funds a Sentiment Index?," *Journal of Finance*, 48(2), 795–800.

CHEN, R.-R. (1995): "A Two-Factor, Preference-Free Model for Interest Rate Sensitive Claims," *Journal of Futures Markets*, 15(3), 345–372.

——— (1996b): *Understanding and Managing Interest Rate Risk.* Singapore: World Scientific.

CHEN, R.-R., AND L. SCOTT (1992): "Pricing Interest Rate Options in a Two-Factor Cox-Ingersoll-Ross Model of the Term Structure," *Review of Financial Studies*, 5(4), 613–636.

——— (1993): "Maximum Likelihood Estimation for a Multifactor Equilibrium Model of the Term Structure of Interest Rates," *Journal of Fixed Income*, 3(3), 14–31.

——— (2003): "Multi-Factor Cox-Ingersoll-Ross Models of the Term Structure: Estimates and Tests from a State-Space Model using a Kalman Filter," *Journal of Real Estate Finance and Economics*, 27(2), 143–172.

CHORDIA, T., AND B. SWAMINATHAN (1996): "Market Segmentation, Imperfect Information, and Closed-End Fund Discounts," Working paper, Vanderbilt University.

CLEWLOW, L., AND C. STRICKLAND (1999): "Valuing Energy Options in a One Factor Model Fitted to Forward Prices," Working paper, Warwick Business School.

COLLIN-DUFRESNE, P., AND R. S. GOLDSTEIN (2001): "The Determinants of Credit Spread Changes," *Journal of Finance*, 56(6), 2177–2207.

——— (2003): "Generalizing the Affine Framework to HJM and Random Field Models," Working paper, Carnegie Mellon University.

CORTAZAR, G., AND L. NARANJO (2003): "A Multi-Factor Stochastic Model and Estimation Procedure for the Valutation and Hedging of Commodity Contingent Claims," *Forthcoming: European Journal of Operations Research*, Working Paper, Pontificia Universidad Catolica de Chile.

CORTAZAR, G., L. NARANJO, AND E. S. SCHWARTZ (2003): "Term Structure Estimation in Low-Frequency Transaction Markets: A Kalman Filter Approach with Incomplete Panel-Data," Working paper, University of California, Los Angeles.

CORTAZAR, G., AND E. S. SCHWARTZ (1994): "The Valuation of Commodity-Contingent Claims," *Journal of Derivatives*, 1(4), 27–39.

——— (2003): "Implementing a Stochastic Model for Oil Futues Prices," *Energy Journal*, 25(3), 215–238.

COURTADON, G. (1982): "The Pricing of Options on Default-Free Bonds," *Journal of Financial and Quantitative Analysis*, 17, 75–100.

COX, J. C., J. E. INGERSOLL, AND S. A. ROSS (1981a): "A Re-Examination of Traditional Hypotheses about the Term Structure of Interest Rates," *Journal of Finance*, 36(4), 769–799.

——— (1981b): "The Relation Between Forward Prices and Futures Prices," *Journal of Financial Economics*, 9, 321–346.

——— (1985a): "An Intertemporal General Equilibrium Model of Asset Prices," *Econometrica*, 53(2), 363–384.

——— (1985b): "A Theory of the Term Structure of Interest Rates," *Econometrica*, 53(2), 385–407.

Crouhy, M., D. Galai, and R. Mark (2000): "A Comparative Analysis of Current Credit Risk Models," *Journal of Banking and Finance*, 24, 59–117.

Dai, Q., and K. J. Singleton (2000): "Specification Analysis of Affine Term Structure Models," *Journal of Finance*, 55(5), 1943–1978.

——— (2002): "Expectation Puzzles, Time-Varying Risk Premia, and Affine Models of the Term Structure," *Journal of Financial Economics*, 63, 415–441.

——— (2003a): "Fixed-Income Pricing," Working paper, New York University.

——— (2003b): "Term Structure Dynamics in Theory and Reality," *Review of Financial Studies*, 16(3), 631–678.

Das, S. (2004): *Swaps and Financial Derivatives: Products, Pricing, Applications and Risk Management.* Chichester: John Whiley & Sons, 3 edn.

De Jong, F. (2000): "Time-Series and Cross-Section Information in Affine Term Structure Models," *Journal of Business & Economic Statistics*, 18(3), 300–314.

De Jong, F., and P. Santa-Clara (1999): "The Dynamics of the Forward Interest Rate Curve: A Formulation with State Variables," *Journal of Financial and Quantitative Analysis*, 34(1), 131–157.

De Long, J. B., A. Shleifer, L. H. Summers, and R. J. Waldmann (1990): "Noise Trader Risk in Financial Markets," *Journal of Political Economy*, 98(4), 703–738.

Deaton, A., and G. Laroque (1992): "On the Behavior of Commodity Prices," *Review of Economic Studies*, 59, 1–23.

——— (1996): "Competitive Storage and Commodity Price Dynamics," *Journal of Political Economy*, 104(5), 896–923.

Delbaen, F., and W. Schachermayer (1994): "A General Version of the Fundamental Theorem of Asset Pricing," *Mathematische Annalen*, 300, 463–520.

DeMunnik, J. F., and P. C. Schotman (1994): "Cross-Sectional versus Time Series Estimation of Term Structure Models: Empirical Results for the Dutch Bond Market," *Journal of Banking and Finance*, 18, 997–1025.

Deng, S. (2000): "Stochastic Models of Energy Commodity Prices and Their Applications: Mean-Reversion with Jumps and Spikes," Working paper, University of California Energy Institute.

Dennis, J. E., and R. B. Schnabel (1996): *Numerical Methods for Unconstrained Optimization and Nonlinear Equations.* Philadelphia: SIAM.

Dimson, E., and C. Minio-Kozerski (1999): "Closed-End Funds: A Survey," *Financial Markets, Institutions & Instruments*, 8(2), 1–41.

Diwan, I., V. R. Errunza, and L. W. Senbet (1995): "The Pricing of Country Funds from Emerging Markets: Theory and Evidence," Working paper, University of Maryland.

Dothan, M. U. (1990): *Prices in Financial Markets.* New York: Oxford University Press.

Duan, J.-C. (1994): "Maximum Likelihood Estimation Using Price Data of the Derivative Contract," *Mathematical Finance*, 4(2), 155–167.

Duan, J.-C., and J.-G. Simonato (1999): "Estimating and Testing Exponential-Affine Term Structure Models by Kalman Filter," *Review of Quantitative Finance and Accounting*, 13, 111–135.

Duarte, J. (2003): "Evaluating an Alternative Risk Preference in Affine Term Structure Models," *Forthcoming: Review of Financial Studies*, Working Paper, University of Washington.

Duffee, G. R. (2002): "Term Premia and Interest Rate Forecasts in Affine Models," *Journal of Finance*, 57, 405–443.

Duffie, D. (1989): *Futures Markets.* Englewood Cliffs: Prentice Hall.

——— (2001): *Dynamic Asset Pricing Theory.* Princeton: Princeton University Press, 3 edn.

Duffie, D., and S. Gray (1995): "Volatility in Energy Prices," in *Managing Energy Price Risk*, ed. by R. Jamson, pp. 39–56. London: Risk Publications.

Duffie, D., and R. Kan (1996): "A Yield-Factor Model of Interest Rates," *Mathematical Finance*, 6(4), 379–406.

Duffie, D., J. Pan, and K. J. Singleton (2000): "Transform Analysis and Asset Pricing for Affine Jump Diffusions," *Econometrica*, 68, 1343–1376.

Duffie, D., and K. J. Singleton (1997): "An Econometric Model of the Term Structure of Interest-Rate Swap Yields," *Journal of Finance*, 52(4), 1287–1321.

Durbin, J., and S. J. Koopman (2001): *Time Series Analysis by State Space Models.* Oxford: Oxford University Press.

Dybvig, P. H., and S. A. Ross (1987): "Arbitrage," in *The New Palgrave: A Dictionary of Economics*, pp. 100–106. New York: Stockton Press.

Engle, R. F., and C. W. J. Granger (1987): "Co-Integration and Error Correction: Representation, Estimation, and Testing," *Econometrica*, 55, 251–276.

Eom, Y. H., J. Helwege, and J.-Z. Huang (2002): "Structural Models of Corporate Bond Pricing: An Empirical Ananlysis," Working paper, Penn State University College of Business.

Escribano, A., J. I. Pena, and P. Villaplana (2002): "Modeling Electricity Prices: International Evidence," Working paper, Universidad Carlos III de Madrid.

Eydeland, A., and H. Geman (1998): "Pricing Power Derivatives," *Risk*, 10, 71–73.

Fama, E. F. (1965): "The Behaviour of Stock Market Prices," *Journal of Business*, 38, 34–104.

——— (1970): "Efficient Capital Markets: A Review of Theory and Empirical Work," *Journal of Finance*, 25, 383–417.

——— (1991): "Efficient Capital Markets: II," *Journal of Finance*, 46(5), 1575–1617.

FARKAS, J. (1902): "Theorie der einfachen Ungleichungen," *Journal für die Reine und Angewandte Mathematik*, 124(1), 1–27.

FEYNMAN, R. P. (1948): "Space-Time Approach to Non-Relativistic Quantum Mechanics," *Reviews of Modern Physics*, 20(2), 367–387.

GALLANT, A. R., AND H. WHITE (1988): *A Unified Theory of Estimation and Inference for Nonlinear Dynamic Models*. Oxford: Blackwell.

GASBARRO, D., R. D. JOHNSON, AND J. K. ZUMWALT (2003): "Evidence on the Mean-Reverting Tendencies of Closed-End Fund Discounts," *Financial Review*, 38, 273–291.

GEMAN, H. (1989): "The Importance of the Forward Neutral Probability in a Stochastic Approach of Interest Rates," Working paper, ESSEC.

GEMAN, H., N. E. KAROUI, AND J.-C. ROCHET (1995): "Changes of Numeraire, Changes of Probability Measure, and Option Pricing," *Journal of Applied Probability Trust*, 32(2), 443–458.

GEMMILL, G., AND D. C. THOMAS (2002): "Noise Trading, Costly Arbitrage, and Asset Pricing: Evidence from Closed-End Funds," *Journal of Finance*, 57(6), 2571–2594.

GEYER, A. L. J., AND S. PICHLER (1999): "A State-Space Approach to Estimate and Test Multi-Factor Cox-Ingersoll-Ross Models of the Term Structure," *Journal of Financial Research*, 22(1), 107–130.

GIBBONS, M. R., AND K. RAMASWAMY (1993): "A Test of the Cox, Ingersoll, and Ross Model of the Term Structure," *Review of Financial Studies*, 6(3), 619–658.

GIBSON, R., AND E. S. SCHWARTZ (1990): "Stochastic Convenience Yield and the Pricing of Oil Contingent Claims," *Journal of Finance*, 45(3), 959–976.

GIRSANOV, I. V. (1960): "On Transforming a Certain Class of Stochastic Processes by Absolutely Continuous Substitution of Measures," *Theory of Probability and its Applications*, 5(3), 285–301.

GOLDSTEIN, R. (2000): "The Term Structure of Interest Rates as a Random Field," *Review of Financial Studies*, 13, 365–384.

GOLUB, G. H., AND C. F. VAN LOAN (1996): *Matrix Computations*. Baltimore: Johns Hopkins Press, 3 edn.

GOURIEROUX, C., AND A. MONFORT (1995a): *Statistics and Econometric Models*, vol. 2. Cambridge: Cambridge University Press.

——— (1995b): *Statistics and Econometric Models*, vol. 1. Cambridge: Cambridge University Press.

——— (1997): *Time Series and Dynamic Models*. Cambridge: Cambridge University Press.

GOURIEROUX, C., A. MONFORT, E. RENAULT, AND A. TROGNON (1984): "Pseudo Maximum Likelihood Methods: Theory," *Econometrica*, 52(3), 681–721.

GREEN, T. C., AND S. FIGLEWSKI (1999): "Market Risk and Model Risk for a Financial Institution Writing Options," *Journal of Finance*, 54(4), 1465–1499.

GRINBLATT, M., AND F. A. SCHWARTZ (2000): "Financial Innovation and the Role of Derivative Securities: An Empirical Analysis of the Treasury STRIPS Program," *Journal of Finance*, 55(3), 1415–1436.

HAMILTON, J. D. (1994a): "State-Space Models," in *Handbook of Econometrics*, ed. by R. Engle, and D. McFadden, vol. 4, pp. 3041–3068. Amsterdam: North-Holland.

——— (1994b): *Time Series Analysis*. Princeton: University Press.

HAN, B. (2003): "Stochastic Volatilities and Correlations of Bond Yields," Working paper, Ohio State University.

HANSEN, P. R. (1998): "Structural Breaks in the Term Structure of Interest Rates," Working paper, University of California, San Diego.

HARRISON, J. M., AND D. M. KREPS (1979): "Martingales and Arbitrage in Multiperiod Securities Markets," *Journal of Economic Theory*, 20, 381–408.

HARRISON, J. M., AND S. R. PLISKA (1981): "Martingales and Stochastic Integrals in the Theory of Continuous Trading," *Stochastic Processes and Their Applications*, 11(1), 215–260.

HARVEY, A. C. (1989): *Forecasting, Structural Time Series Models, and the Kalman Filter*. Cambridge: Cambridge University Press.

——— (1993): *Time Series Models*. New York: Harvester Wheatsheaf, 2 edn.

HEATH, D., R. JARROW, AND A. MORTON (1992): "Bond Pricing and the Term Structure of Interest Rates: A New Methodology for Contingent Claims Valuation," *Econometrica*, 60(1), 77–105.

HESTON, S. L. (1992): "Testing Continuous Time Models of the Term Structure of Interest Rates," Working paper, Yale University.

——— (1993): "A Closed-Form Solution for Options with Stochastic Volatility with Applications to Bond and Currency Options," *Review of Financial Studies*, 6(2), 327–343.

HICKS, J. (1939): *Value and Capital*. London: Oxford University Press.

HILLIARD, J. E., AND J. REIS (1998): "Valuation of Commodity Futures and Options under Stochastic Convenience Yields, Interest Rates, and Jump Diffusions in the Spot," *Journal of Financial and Quantitative Analysis*, 33(1), 61–86.

HIRSCH, M. W., AND S. SMALE (1974): *Differential Equations, Dynamical Systems, and Linear Algebra*. San Diego: Academic Press.

HO, T. S., AND S.-B. LEE (1986): "Term Structure Movements and Pricing Interest Rate Contingent Claims," *Journal of Finance*, 41(5), 1011–1029.

HO, T. S. Y. (1992): "Key Rate Durations: Measures of Interest Rate Risk," *Journal of Fixed Income*, 2(2), 29–44.

HUANG, J.-Z., AND M. HUANG (2003): "How much of the Corporate-Treasury Yield Spread is Due to Credit Risk?," Working paper, Stanford University School of Business.

HULL, J., AND A. WHITE (1987): "The Pricing of Options on Assets with Stochastic Volatilities," *The Journal of Finance*, 42(2), 281–300.

HULL, J. C. (2003): *Options, Futures, and Other Derivatives*. Englewood Cliffs: Prentice Hall, 5 edn.

HULL, J. C., AND A. WHITE (1990a): "Pricing Interest-Rate-Derivative Securities," *Review of Financial Studies*, 3(4), 573–592.

——— (1990b): "Valuing Derivative Securities Using the Explicit Finite Difference Method," *Journal of Financial and Quantitative Analysis*, 25(1), 87–100.

INGERSOLL, J. E. (1976): "A Theoretical and Empirical Investigation of Dual Purpose Funds," *Journal of Financial Economics*, 3, 83–123.

——— (1987): *Theory of Financial Decision Making*. Savage: Rowman & Littlefield.

JAMESON, R. (1997): *The US Power Market*. London: Risk Publications.

JAMSHIDIAN, F. (1989): "An Exact Bond Option Formula," *Journal of Finance*, 44(1), 205–209.

JAMSHIDIAN, F. (1997): "Libor and Swap Market Models and Measures," *Finance and Stochastics*, 1, 293–330.

JARROW, R. A., AND G. S. OLDFIELD (1981): "Forward Contracts and Futures Contracts," *Journal of Financial Economics*, 9, 373–382.

JAZWINSKI, A. H. (1970): *Stochastic Processes and Filtering Theory*. New York: Academic Press.

JOHNSON, B., AND A. SOGOMONIAN (1997): "Electricity Futures," in *The US Power Market*, ed. by R. Jameson, pp. 83–98. London: Risk Publications.

JOLLIFFE, I. T. (1986): *Principal Component Analysis*. New York: Springer.

KAC, M. (1949): "On the Distribution of Certain Wiener Functionals," *Transactions of the American Mathematical Society*, 65, 1–13.

KALDOR, N. (1939): "Speculation and Economic Stability," *Review of Economic Studies*, 7, 1–27.

KALMAN, R. E. (1960): "A New Approach to Linear Filtering and Prediction Problems," *Journal of Basic Engineering*, 82, 35–45.

——— (1963): "New Methods in Wiener Filtering Theory," in *Proceedings of the First Symposium of Engineering Applications of Random Function Theory and Probability*, ed. by J. L. Bogdanoff, and F. Kozin, pp. 270–388. New York: Whiley.

KALMAN, R. E., AND R. S. BUCY (1961): "New Results in Linear Filtering and Prediction Theory," *Journal of Basic Engineering*, 83, 95–108.

KAMINSKI, V. (1997): "The Challenge of Pricing and Risk Managing Electricity Derivatives," in *The US Power Market*, ed. by R. Jameson. London: Risk Publications.

KAMKE, E. (1979): *Differentialgleichungen - Lösungsmethoden und Lösungen*, vol. II. Stuttgart: Teubner, 6 edn.

——— (1983): *Differentialgleichungen - Lösungsmethoden und Lösungen*, vol. I. Stuttgart: Teubner, 10 edn.

KARATZAS, I., AND S. E. SHREVE (1991): *Brownian Motion and Stochastic Calculus*. Berlin: Springer, 2 edn.

KARESEN, K. F., AND E. HUSBY (2002): "A Joint State-Space Model for Electricity Spot and Futures Prices," Working paper, Norwegian Computing Center.

KARLIN, S., AND H. M. TAYLOR (1975): *A First Course in Stochastic Processes*. Boston: Academic Press, 2 edn.

KELLERHALS, B. P. (2001): *Financial Pricing Models in Continuous Time and Kalman Filtering*. New York: Springer.

KELLERHALS, B. P., AND M. UHRIG-HOMBURG (1998): "Ungleichgewichte auf Bondmärkten: Aktive Handelsstrategien auf Basis geschätzter Zinsstrukturkurven," *Finanzmarkt und Portfolio Management*, 12(1), 32–45.

KENNEDY, D. P. (1994): "The Term Structure of Interest Rates as a Gaussian Random Field," *Mathematical Finance*, 4(3), 247–258.

——— (1997): "Characterizing Gaussian Models of the Term Structure of Interest Rates," *Mathematical Finance*, 7(2), 107–118.

KERKHOF, J., AND A. PELSSER (2002): "Observational Equivalence of Discrete String Models and Market Models," *Journal of Derivatives*, 10, 55–61.

KHORANA, A., S. WAHAL, AND M. ZENNER (2002): "Agency Conflicts in Closed-End Funds: The Case of Rights Offerings," *Journal of Financial and Quantitative Analysis*, 37(2), 177–200.

KIMMEL, R. L. (2003): "Modeling the Term Structure of Interest Rates: A New Approach," *Forthcoming: Journal of Financial Economics*, Working Paper, Princeton University.

KLAFFKY, T. E., Y. Y. MA, AND A. NOZARI (1992): "Managing Yield Curve Exposure: Introducing Reshaping Durations," *Journal of Fixed Income*, 2(3), 39–45.

KNITTEL, C. R., AND M. R. ROBERTS (2001): "An Empirical Examination of Deregulaated Electricity Prices," Working paper, Boston University School of Mangement.

KYLE, A. S. (1985): "Continuous Auctions and Insider Trading," *Econometrica*, 53, 1315–1335.

LEE, C. M., A. SHLEIFER, AND R. H. THALER (1990): "Anomalies: Closed-End Mutual Funds," *Journal of Economic Perspectives*, 4(4), 153–164.

——— (1991): "Investor Sentiment and the Closed-End Fund Puzzle," *Journal of Finance*, 46(1), 75–109.

LEONG, K. S. (1997): "The Forward Curve in the Electricity Market," in *The US Power Market*, ed. by R. Jameson, pp. 133–148. London: Risk Publications.

LITTERMAN, R., AND J. SCHEINKMAN (1991): "Common Factors Affecting Bond Returns," *Journal of Fixed Income*, 1(6), 54–61.

LITTERMAN, R., J. SCHEINKMAN, AND L. WEISS (1991): "Volatility and the Yield Curve," *Journal of Fixed Income*, 1(6), 49–53.

LONGSTAFF, F. A., P. SANTA-CLARA, AND E. S. SCHWARTZ (2001a): "The Relative Valuation of Caps and Swaptions: Theory and Empirical Evidence," *Journal of Finance*, 56, 2067–2109.

——— (2001b): "Throwing away a Billion Dollars: The Cost of Suboptimal Exercise Strategies in the Swaption Markets," *Journal of Financial Economics*, 62, 39–66.

LONGSTAFF, F. A., AND E. S. SCHWARTZ (1992): "Interest Rate Volatility and the Term Structure: A Two-Factor General Equilibrium Model," *Journal of Finance*, 47(4), 1259–1282.

LONGSTAFF, F. A., AND A. W. WANG (2002): "Electricity Forward Prices: A High Frequency Empirical Analysis," Working paper, University of California, Los Angeles.

——— (2003): "An Empirical Analysis of the Risk Premia in Electricity Forward Prices," Working paper, University of California, Los Angeles.

LOVE, D. A., AND S. F. MCSHEA (1997): "The U.S. Treasury STRIPS: Valuation, Liability, and Asset Allocation Frontier," in *Yield Curve Dynamics*, ed. by R. J. Ryan, pp. 9–32. Chicago: Glenlake.

LUCIA, J. J., AND E. S. SCHWARTZ (2002): "Electricity Prices and Power Derivatives: Evidence from the Nordic Power Exchange," *Review of Derivatives Research*, 5, 5–50.

LUND, J. (1997): "Five Essays in Financial Econometrics and the Term Structure of Interest Rates," Phd dissertation, Aarhus School of Business.

LÜTKEPOHL, H. (1996): *Handbook of Matrices.* Chichester: John Whiley & Sons.

MACAULAY, F. R. (1938): *Some Theoretical Problems Suggested by the Movements of Interest Rates, Bond Yields, and Stock Prices in the U.S. Since 1856.* New York: National Bureau of Economic Research.

MALKIEL, B. G. (1977): "The Valuation of Closed-End Investment-Company Shares," *Journal of Finance*, 32(3), 847–859.

MANDELBROT, B. (1963): "The Variation of Certain Speculative Prices," *Journal of Business*, 36, 394–419.

MANOLIU, M., AND S. TOMPAIDIS (2002): "Energy Futures Prices: Term Strucutre Models with Kalman Filter Estimation," *Applied Mathematical Finance*, 9(1), 21–43.

MARKHAM, J. W. (1987): *The History of Commodity Futures Trading and Its Regulation.* New York: Praeger.

MARSHALL, J. F., AND K. R. KAPNER (1993): *The Swaps Market.* Miami: Kolb, 2 edn.

MCCULLOCH, J. H. (1975): "The Tax-Adjusted Yield Curve," *Journal of Finance*, 30(3), 811–830.

MCCULLOCH, J. H., AND H.-C. KWON (1993): "U.S. Term Structure Data. 1947-1991," Working paper, Ohio State University.

MCDONALD, J. B. (1996): "Probability Distributions for Financial Models," in *Handbook of Statistics*, ed. by G. Maddala, and C. R. Rao, vol. 14, pp. 427–461. Amsterdam: Elsevier.

MERTON, R. C. (1973): "Theory of Rational Option Pricing," *Bell Journal of Economics and Management Science*, 4, 141–183.

——— (1976): "Option Pricing when Underlying Stock Prices are Discontinuous," *Journal of Financial Economics*, 3, 125–144.

——— (1992): *Continuous-Time Finance.* Cambridge: Blackwell, rev. edn.

MILLER, R. L., AND D. D. VANHOOSE (1993): *Modern Money and Banking.* New York: McGraw-Hill, 3 edn.

MILTERSEN, K. R., K. SANDMANN, AND D. SONDERMANN (1997): "Closed Form Solutions for Term Structure Derivatives with Log-Normal Interest Rates," *Journal of Finance*, 52(1), 409–430.

MILTERSEN, K. R., AND E. S. SCHWARTZ (1998): "Pricing of Options on Commodity Futures with Stochastic Term Structures of Convenience Yields and Interest Rates," *Journal of Finance and Quantitative Analysis*, 33(1), 33–59.

MINIO-PALUELLO, C. (1998): "The UK Closed-End Fund Discount," Phd thesis, London Business School.

MODIGLIANI, F., AND R. SUTCH (1966): "Innovations in Interest Rate Policy," *American Economic Review*, 56, 178–97.

MOLER, C., AND C. VAN LOAN (1978): "Nineteen Dubious Ways to Compute the Exponential of a Matrix," *SIAM Review*, 20(4), 801–836.

MORTON, A. J. (1988): "Arbitrage and Martingales," Technical report, Cornell University.

MUSIELA, M., AND M. RUTKOWSKI (1997): *Martingale Methods in Financial Modelling.* Berlin: Springer.

NUNES, J. P. V., AND L. CLEWLOW (1999): "Kalman Filtering of Gaussian Exponential-Affine LIBOR-Rate Models Using Caps and Swaptions," Working paper, University of Warwick.

OKSENDAL, B. (2000): *Stochastic Differential Equations.* Berlin: Springer, 5 edn.

PEARSON, N. D., AND T.-S. SUN (1994): "Exploiting Conditional Density in Estimating the Term Structure: An Application to the Cox, Ingersoll, and Ross Model," *Journal of Finance*, 49(4), 1279–1304.

PEAVY, J. W. (1990): "Returns on Initial Public Offerings of Closed-End Funds," *Review of Financial Studies*, 3(4), 695–708.

PENNACCHI, G. G. (1991): "Identifying the Dynamics of Real Interest Rates and Inflation: Evidence Using Survey Data," *The Review of Financial Studies*, 4(1), 53–86.

PHILLIPS, P. C. B., AND S. OULIARIS (1988): "Testing for Cointegration Using Principal Component Methods," *Journal of Economic Dynamics and Control*, 12, 205–230.

PIAZZESI, M. (2003): "Affine Term Structure Models," Working paper, University of California, Los Angeles.

PILIPOVIC, D. (1998): *Energy Risk - Valuing and Managing Energy Risk.* New York: McGraw-Hill.

PINDYCK, R. S. (1999): "The Long-Run Evolution of Energy Prices," *Energy Journal*, 20(2), 1–27.

PIRRONG, C., AND M. JERMAKYAN (2000): "The Price of Power - The Valuation of Power and Weather Derivatives," Working paper, Olin School of Business, Washington University.

POKALSKY, J., AND J. ROBINSON (1997): "Integrating Physical and Financial OTC Contract Portfolios," in *The US Power Market*, ed. by R. Jameson, pp. 111–132. London: Risk Publications.

PONTIFF, J. (1995): "Closend-End Fund Premia and Returns Implications for Financial Market Equilibrium," *Journal of Financial Economics*, 37, 341–370.

——— (1997): "Excess Volatility and Closed-End Funds," *The American Economic Review*, 87, 155–169.

PRESS, W. H., S. TEUKOLSKY, W. VETTERLING, AND B. FLANNERY (2002): *Numerical Recipes in C++: The Art of Scientific Computing*. Cambridge: Cambridge University Press, 2 edn.

REBONATO, R. (1998): *Interest-Rate Option Models*. Chichester: John Whiley & Sons, 2 edn.

RICHARD, S. F., AND M. SUNDARESAN (1981): "A Continuous Time Equilibrium Model of Forward Prices and Futures Prices in a Multigood Economy," *Journal of Financial Economics*, 9, 347–371.

RICHARDS, R. M., D. R. FRASER, AND J. C. GROTH (1980): "Winning Strategies for Closed-End Funds," *Journal of Portfolio Management*, 7(1), 50–55.

ROENFELDT, R. L., AND D. L. TUTTLE (1973): "An Examination of the Discounts and Premiums of Closed-End Investment Companies," *Journal of Business Research*, 2(1), 129–140.

ROGERS, L. (1995): "Which Model for the Term Structure of Interest Rates Should One Use?," in *Mathematical Finance*, ed. by M. H. Davis, D. Duffie, W. H. Fleming, and S. E. Shreve, pp. 93–115. Berlin: Springer Verlag.

ROSS, S. A. (1997): "Hedging Long-Run Commitments: Exercises in Incomplete Market Pricing," *Economic Notes by Banca Monte die Paschi di Siena SpA*, 26(2), 385–420.

ROTHENBERG, T. J. (1971): "Identification in Parametric Models," *Econometrica*, 39(3), 577–591.

ROUTLEDGE, B. R., D. J. SEPPI, AND C. S. SPATT (2000): "Equilibrium Forward Curves of Commodities," *Journal of Finance*, 55(3), 1297–1338.

——— (2001): "The 'Spark Spread': An Equilibrium Model of Cross-Commodity Price Relationships in Electricity," Working paper, Carnegie Mellon University.

SANTA-CLARA, P., AND D. SORNETTE (2001): "The Dynamics of the Forward Interest Rate and Curve with Stochastic String Shocks," *Review of Financial Studies*, 14, 149–185.

SCHAEFER, S. M., AND E. S. SCHWARTZ (1984): "A Two-Factor Model of the Term Structure: An Approximate Analytical Solution," *Journal of Financial and Quantitative Analysis*, 19(4), 413–424.

SCHÖBEL, R. (1992): "Arbitrage Pricing in Commodity Futures Markets: Is It Feasible?," Presentation, Conference on Mathematical Finance, Oberwolfach, Germany, 23.08.-29.08.1992.

——— (1995a): "Inflation and Valuation: A Gaussian Model of the Term Structure of Interest Rates," Working paper, Universität Tübingen.

——— (1995b): *Kapitalmarkt und zeitkontinuierliche Bewertung*. Berlin: Springer.

SCHÖBEL, R., AND J. ZHU (1999): "Stochastic Volatility with an Ornstein-Uhlenbeck Process: An Extension," *European Finance Review*, 3, 23–46.

SCHOENBUCHER, P. J. (2003): *Credit Derivatives Pricing Models - Models, Pricing and Implementation.* Chichester: John Whiley & Sons.

SCHWARTZ, E., AND J. E. SMITH (2000): "Short-Term Variations and Long-Term Dynamics in Commodity Prices," *Management Science*, 46(7), 893–911.

SCHWARTZ, E. S. (1977): "The Valuation of Warrants: Implementating a New Approach," *Journal of Financial Economics*, 4, 79–94.

——— (1997): "The Stochastic Behaviour of Commodity Prices: Implications for Valuation and Hedging," *Journal of Finance*, 52(3), 923–973.

SCHWEPPE, F. C. (1965): "Evaluation of Likelihood Functions for Gaussian Signals," *IEEE Transactions on Information Theory*, 11, 61–70.

SCOTT, L. O. (1997): "Pricing Stock Options in a Jump-Diffusion Model with Stochastic Volatility and Interest Rates: Applications of Fourier Inversion Methods," *Mathematical Finance*, 7(4), 413–426.

SHARPE, W. F., AND H. B. SOSIN (1975): "Closed-End Investment Companies in the United States," in *European Finance Association: 1974 Proceedings*, ed. by B. Jacquillat, pp. 37–63. Amsterdam: North-Holland.

SMITH, C. W. J. (1976): "Option Pricing - A Review," *Journal of Financial Economics*, 3, 3–51.

SORENSEN, C. (2002): "Modeling Seasonality in Agricultural Commodity Futures," *Journal of Futures Markets*, 22(5), 393–426.

SPIEGEL, M. (1999): "Closed-End Fund Discounts in a Rational Agent Economy," Working paper, Yale School of Management.

STEIN, E. M., AND J. C. STEIN (1991): "Stock Price Distributions with Stochastic Volatility: An Analytic Approach," *Review of Financial Studies*, 4(4), 727–752.

STOLL, H. R. (1969): "The Relationship Between Put and Call Option Prices," *Journal of Finance*, 14(5), 801–824.

STOLL, H. R., AND R. E. WHALEY (1993): *Futures and Options - Theory and Applications.* Cincinnati: South-Western Publishing.

STRICKER, C. (1984): "Integral Representation in the Theory of Continuous Trading," *Stochastics*, 13, 249–257.

STUART, A., AND J. K. ORD (1991): *Kendall's Advanced Theory of Statistics*, vol. 2. London: Edward Arnold, 5 edn.

——— (1994): *Kendall's Advanced Theory of Statistics*, vol. 1. London: Edward Arnold, 6 edn.

SUNDARESAN, S. M. (2000): "Continuous-Time Methods in Finance: A Review and an Assessment," *Journal of Finance*, 55(4), 1569–1622.

SWAMINATHAN, B. (1996): "Time-Varying Expected Small Firm Returns and Closed-End Fund Discounts," *Review of Financial Studies*, 9(3), 845–887.

TANIZAKI, H. (1996): *Nonlinear Filters.* Berlin: Springer, 2 edn.

THEIL, H. (1971): *Principles of Econometrics.* Amsterdam: North-Holland.

THOMPSON, R. (1978): "The Information Content of Discounts and Premiums on Closed-End Fund Shares," *Journal of Financial Economics*, 6, 151–186.

TITMAN, S., AND W. TOROUS (1989): "Valuing Commercial Mortgages: An Empirical Investigation of the Contingent-Claims Approach to Pricing Risky-Debt," *Journal of Finance*, 44(2), 345–373.

UHRIG, M., AND U. WALTER (1996): "A New Numerical Approach for Fitting the Initial Yield Curve," *Journal of Fixed Income*, 6(3), 82–90.

UHRIG-HOMBURG, M. (1999): "Die Bedeutung der Mean-Reversion von Zinsprozessen für Optionswerte: Das Beispiel der Korridor-Zinsoption," *OR Spektrum*, 21, 183–203.

UN (1996): "Standard International Trade Classification (Revision 3)," Statistical Papers 34, New York: United Nations Publications.

VARIAN, H. R. (1987): "The Arbitrage Principle in Financial Economics," *Journal of Economic Perspectives*, 1(2), 55–72.

VASICEK, O. A. (1977): "An Equilibrium Characterization of the Term Structure," *Journal of Financial Economics*, 5(2), 177–188.

WALDMAN, M. (1992): "Beyond Duration: Risk Dimensions of Mortgage Securities," *Journal of Fixed Income*, 2(12), 5–15.

WEISS, K. (1989): "The Post-Offering Price Performance of Closed-End Funds," *Financial Management*, 18, 57–67.

WHITE, H. (1982): "Maximum Likelihood Estimation of Misspecified Models," *Econometrica*, 50(1), 1–25.

——— (1994): *Estimation, Inference, and Specification Analysis*. Cambridge: Cambridge University Press.

WILMOTT, P., J. DEWYNNE, AND S. HOWISON (1993): *Option Pricing: Mathematical Models and Computation*. Oxford: Oxford Financial Press.

WOLAK, F. A. (1997): "Market Design and Price Behavior in Restructured Electricity Markets - An International Comparison," Working paper, University of California Energy Institute.

WOODLEY, J. A. C., AND H. L. HUNT (1997): "Reinvesting the Electricity Industry: Implications of an Electricity Commodity Market," in *The US Power Market*, ed. by R. Jameson, pp. 43–55. London: Risk Publications.

WORKING, H. (1949): "The Theory of the Price of Storage," *American Economic Review*, 39, 1254–1262.

WU, X. (2000): "A New Stochastic Duration Based on the Vasicek and CIR Term Structure Theories," *Management Science*, 27(7&8), 911–932.

ZHU, J. (2000): *Modular Pricing of Options - An Application of Fourier Analysis*. New York: Springer.

Index

Druck: betz-druck GmbH, D-64291 Darmstadt
Verarbeitung: Buchbinderei Schäffer, D-67269 Grünstadt